HOMO

Abhijit Naskar is one of the world's renowned Neuroscientists, International Bestselling Author of "The God Parasite: Revelation of Neuroscience", who has taught the world with his scientific works how the basic awareness of Neurobiology can completely redefine our perception of life and make our daily life much more cheerful and meaningful. His books on neuroscience have built a bridge between the layman's life and modern neuroscience.

Also by Abhijit Naskar

The Art of Neuroscience in Everything

Your Own Neuron: A Tour of Your Psychic Brain

The God Parasite: Revelation of Neuroscience

The Spirituality Engine

Love Sutra: The Neuroscientific Manual of Love

Neurosutra: The Abhijit Naskar Collection

Autobiography of God: Biopsy of A Cognitive Reality

Biopsy of Religions: Neuroanalysis towards Universal Tolerance

Prescription: Treating India's Soul

What is Mind?

HOMO
A BRIEF HISTORY OF CONSCIOUSNESS

(Revised Edition)

ABHIJIT NASKAR

Copyright © 2015 Abhijit Naskar

This is a work of non-fiction

First Published in United States of America, in 2015

All rights reserved. No part of this publication may be reproduced, distributed, or transmitted in any form or by any means, including photocopying, recording, or other electronic or mechanical methods, without the prior written permission of the author, except in the case of brief quotations embodied in critical reviews and certain other noncommercial uses permitted by copyright law.

An Amazon Publishing Company, 2nd Edition, 2017

Printed in United States of America

ISBN-13: 978-1544895666

Dedicated to Sir Roger Penrose

Contents

Foreword 1

Preface 7

Darwinism 10

Meet Our Ancestors 25

Evolution of Human Brain 60

Evolution of Speech 72

Nature of Consciousness 80

Qualia & Human Reality 91

Bibliography 119

Foreword

Consciousness is with no doubt the most important phenomenon that "we can experiment". I wanted to say "we can observe", but that would not have been appropriate, because precisely we cannot observe it. We feel it from the inside, but up to this day we have no way to measure (from the outside) whether some entity is conscious or not. Obviously we can have some indications helpful for anesthesiologists, but we cannot observe or measure consciousness itself.

It was believed, some time ago, that consciousness was associated with so called EEG waves of the brain, we now know it is not correct. We understand that the whole brain participates to generate conscious states, these states cannot be "located", and they imply long distance synchronizations of various brain regions. Most neuroscientists such as Abhijit Naskar do think that the brain alone is the source of consciousness, whereas rare others believe that the brain acts like a sort of radio receptor and that at least some aspects of consciousness are received from outside. How would that work?

So, isn't that strange that what I have nominated as "the most important phenomenon" (for us humans) in this universe is not even observable or measurable and furthermore it is not clearly defined either (not to say indefinable). It is probably this very difficulty that has lead science to put consciousness aside and neglect it for a long time. I do remember 40 years ago, when I was a student, it was considered "not serious" to try and examine consciousness in a scientific perspective. The whole of science (I mean hard science) did not bother about the existence of this "phenomenon" and somehow did not even recognize its existence.

I believe Quantum Physics, that explicitly needs to mention an observer to be able to describe a phenomenon, has progressively brought it back in the realm of science. Certain thinkers like the mathematician Roger Penrose and the anesthesiologist Stuart Hameroff attribute the existence of consciousness to quantum properties.

I have called consciousness the most important phenomenon for a simple and very profound reason: *our knowledge of any other phenomenon transits through our consciousness*. There would be no language and no science without consciousness. One cannot compare it to any other phenomenon that science studies like let's say black holes because there would be no black holes (as we know them) without our consciousness.

Studying consciousness is a bit like studying the tool that reveals us the reality and while using the tool we cannot get fully outside of it to observe it! Whatever we want to observe we must use this tool. A circularity! But does consciousness really exist or is it a sort of illusion? René Descartes, the French mathematician and philosopher famously wrote: "Je pense donc je suis", I think therefore I am.

I cannot doubt about my own consciousness. I cannot believe and proclaim that I am not conscious! Believing it is already a sign that I am conscious! How do I know that another living being is conscious? Well I don't really know, I assume it, I give him the benefit of doubt, because he looks like me, he behaves like me and he thinks like me. But is acting like being conscious sufficient to definitively say consciousness is there?

Many authors have imagined Zombies that would have the complete appearance of normal individuals, but would not be conscious and would have no morality. Is the behavior a sufficient sign that we are facing the real thing? This question is becoming crucial now with the development of Artificial Intelligence, I will try and say why. But let me say first that intelligence cannot exist without understanding and understanding necessitates consciousness.

Alan Turing, a mathematician who conceived the modern digital computer, in his 1936 paper, faced this kind of difficulty. In 1950 he published an article entitled "Computing machinery and intelligence". In this paper he asks the question: *Can machines think?* Unable to define thought (like we are unable to define consciousness) he proposes what he calls an "Imitation game". It is played with three entities, a man (A), a computer (B), and an interrogator (C). The interrogator stays in a room apart. The object of the game for the interrogator is to determine which of the other two is the man and which is the computer by asking questions. This imitation game is now called the Turing test and certain people believe it can help determine whether a computer is intelligent. No computer has up to now managed to fully fool the interrogator, but even if one day your phone manages to pass the Turing test, will it be a proof that it is intelligent, thus conscious? Or will it be more like a Zombie who can imitate consciousness. Is a really good imitation equal to the real thing? Surely not in my view.

As you will read in Naskar's book, even though in the last few decades many important steps towards the understanding of consciousness have been overcome, there is still a lot to do and I believe that it will involve knowledge far beyond neuroscience and will lead us to redefine our complete self image and to the understanding of who we are in this vast universe. You have in your hands a book that in simple words brings the essentials of what we know about consciousness allowing you to take immediate advantage of this knowledge in your personal life and be ready for the future perspectives.

- Ronald Cicurel, Mathematician, PhD
  Lecturer at the EPFL in Lausanne, Switzerland,
  Member of the Brazilian Committee for the Advancement of Science,
  Fellow of the International Institute of Neuroscience of Natal, Brazil

Preface

What is Consciousness? Consciousness is simply the functional expression of the brain. This *complex organic device is the most fascinating structure on earth that creates and thereafter drives consciousness*. Consciousness evolved hand in hand with the evolution of the human brain throughout a time span of six million years. That is the approximate length of time it took for the Homo sapiens brain to develop once our ancestral line diverged from the line which developed into modern chimpanzees and other apes.

Just like teenage life is all about three things - DRAMA, DRAMA and DRAMA, existence of a species is all about EVOLUTION! The rule of our Mother Earth is "Evolve or Vanish". We followed the rule like a mommy's boy and today we are the smartest species on this planet and this entire solar system. However, Mother

Nature remains indifferent, whether we exist or get extinct.

The human species is the only species that has ever asked the question, where do I come from? It is the only species that creates sonnets, symphonies, and literature. It is the only species that has developed mathematical equations and set foot on the moon. No other species on earth is likely to ever match these achievements. The biological structure that has enabled humans to perform these fascinating feats of intelligence is, without question, the human brain. With the advent of consciousness in our brain circuits, for the first time in planet earth's history we became truly human.

The evolution of various cortical brain circuits enabled the human species to be truly conscious about the extraordinary human abilities. That's exactly what we are about to explore. Let's turn the clock back millions of years and embark on an evolutionary journey through history to visualize how humans slowly started to become humans in the truest sense of the term. Let's go back in time and glance at the most fascinating event in planet earth's history, i.e. Evolution of Consciousness.

Get set for a joyride of a Brief History of Consciousness.

DARWINISM

"My time was wasted, as far as the academical studies were concerned, as completely as at Edinburgh and at school"

- Charles Darwin

All species on earth are related to one another like cousins and distant kin in a vast family tree of life. In the path of studying the family tree of life the name that comes up countless times, is Charles Darwin. After all, his expedition on H.M.S. Beagle eventually became the most astounding journey of biological studies.

Charles Darwin is one of the most important minds of science in the last five hundred years. In his book, "On the Origin of Species by Means of Natural Selection" published in 1859, Darwin proposed what is now called the theory of evolution. This book, along with its best-known companion "The Descent of Man and Selection in Relation to Sex", published in 1871, called for a major change in scientific thinking about the origin of life, particularly in the field of biology. Darwin was not the first scientist to propose a theory of evolution nor was he the foremost thinker on the subject in 1859. Darwin was a product of his time.

Figure 1-1: Charles Darwin (Sketch by Abhijit Naskar, 2015)

Bentley Glass, one of the editors of the book "Forerunners of Darwin", calls Darwin's solution "a magnificent synthesis of evidence." When Thomas Huxley first read Darwin's paper, he reported to have exclaimed: *"How stupid of me not to have thought of that."*

"From September 1854 I devoted my whole time to arranging my huge pile of notes, to observing, and to experimenting in relation to the transmutation of species."

> One of the most auspicious statements in Darwin's autobiography

When Darwin conceived the idea of "natural selection" soon after his return from the voyage of the Beagle in 1838, he set himself to the task of streamlining the arguments and collecting supporting evidence. Although his ideas were well known among English biologists, Darwin hesitated to publish them in fear of the reaction of the religious society and the Church, and he continued to accumulate an increasingly impressive body of data consistent with his theory. Darwin's hand was forced when another biologist Alfred Russel Wallace independently conceived the same idea, and mailed it in a concise ten-page letter to Darwin. Darwin quickly wrote up a short outline, and both papers were presented to the Linnaean Society simultaneously in 1858. Had Wallace bypassed Darwin and published first, right now we would be speaking of Wallace's theory of evolution by natural selection, and Darwin would remain a poorly known documenter of Wallace's theory. Darwin, of course, is a household name. It is a pity that Wallace is almost completely unknown except among biologists and historians of science, since he was an equally brilliant scholar and independently came up with the same idea.

However, instead of disputing endlessly over priority, as many of today's scientists do, Darwin and Wallace cheerfully acknowledged each other's contributions and Wallace even wrote a book called Darwinism, advocating what he referred to as "Darwin's" theory of natural selection. Upon hearing about this book, Darwin responded: *"You should not speak of Darwinism for it can as well be called Wallacism."*

Darwin was a very clever scientist. His use of the research of fellow scientists is one of the most fascinating features of his book "The Origin of Species". He quoted from their work even though many of them actually never did support any kind of theory of evolution.

Now, shall we look into some details of Darwin's life?

Charles Robert Darwin was born on 12 February 1809 in Shrewsbury, an ancient market town near the Welsh border in the county of Shropshire, West Midlands. Before Darwin, the town's most famous resident was Robert Clive, the man who led the British conquest of India in the eighteenth century. In the first twenty years of the nineteenth century, the town had a population of approximately 16,000. Although Shrewsbury had a comparatively small population, it was the county town of Shropshire, the most rural of the English counties. The rural character of the county and Shrewsbury's importance were two of the reasons why Darwin's parents chose to live there. By the time Darwin was born, Darwin's father had a flourishing medical practice that covered the town and the surrounding area. Darwin's interest in and love of nature can be traced to the surroundings of his early childhood.

Apart from the obvious details such as when and where he was born, the first significant detail someone new to his story should know is that the Darwin family and the Darwin name were famous before Charles Darwin was born. Robert Waring Darwin (1766-1848), Darwin's father, had married Susannah (1765-1817), the daughter of Josiah Wedgwood, the famous pottery magnate, in 1796. Darwin had four sisters and one brother: Marianne (1798-1858), Caroline (1800-1888), Susan (1803-1866), Erasmus (1804-1881), and Catherine (1810-1866). Darwin was the fifth child of the six. Darwin grew up in a family whose wealth enabled him to become a gentleman, a man of means who did not need a career to support himself or his future family.

When Darwin was sixteen, he seemed to have little direction and drive in his life. His father made the following caustic comment about him: *"You care for nothing but shooting, dogs, and rat-catching and you will be a disgrace to yourself and all your family"*. The comment is easier to understand if Darwin's family background is taken into account. Robert Darwin had continued the family tradition of brilliance by becoming a fellow of Royal Society at the age of twenty-two and like his father, had a well-respected medical practice. The son of Robert Darwin was privileged in ways that meant much was expected of him.

Darwin became a student of Shrewsbury school at the age of nine. The main subjects taught at the school were the classics, Latin and Greek language and culture. These subjects supposedly turned boys into gentlemen, but Darwin was uninterested in the classics. *"Nothing could have been worse for the development of my mind"*, was Darwin's assessment of his school days.

His father recognized that Darwin needed more purpose in his life and took him out of school two years early. Darwin's elder brother Erasmus was going to Edinburgh University to complete his studies in medicine. Robert Darwin decided that Charles ought to accompany his brother. His plan was for Charles to attend the medical lectures with his brother and when Charles was old enough he too could take the appropriate examinations for his degree.

The plan seemed sensible enough. In the summer before he went to Edinburgh, Darwin had taken care of about twelve of his father's patients. He enjoyed the work and his father thought Darwin would make a successful physician. Counterbalancing this was the fact that Robert Darwin was doing exactly the same as his father Erasmus Darwin. The grandfather forced the father to become a doctor and the father intended to do the same for the son. Robert Darwin submitted, grudgingly, to his father's authority. Charles Darwin, who soon realized that his father would provide financial support sufficient to last his whole life, was less subservient. Darwin found the lectures at Edinburgh *"intolerably dull"* and the sight of human blood made him physically sick. He fled from one particularly stomach-churning operation on a child. The regular use of anesthetics did not occur until the 1850s and his efforts at studying were dilatory at best. Darwin felt no call or inclination to become a doctor when he actually had to study medicine.

Darwin's confessions in letters to his sisters, was bad enough, but Darwin's trips around the countryside to see various friends during the summer of 1827 and his obsession with shooting suggested to Robert Darwin that his son would become a dilettante, a wealthy son squandering his father's money on trivial pursuits. Robert Darwin intervened in his son's life again: *"Charles Darwin would go to Cambridge University and study to become a clergyman"*.

If Robert Darwin believed that his son would settle into a less dissolute life at Cambridge, he was mistaken. Looking back on his years at the university, Darwin claimed that *"my time was wasted, as far as the academic studies were concerned, as completely as at Edinburgh and at school"*. There is no doubt that his father was infuriated by the fact that Darwin continued his shooting, hunting, and riding in the countryside while at Cambridge. Even worse, Darwin also added drinking, jolly singing and playing cards to his leisure activities. He was studying to become a clergyman.

Darwin persuaded himself that he should accept the doctrines of the Church of England fully and did not think too deeply about the literal truth of the Bible or the foundational premises of Paley's arguments. In the end, neither Darwin's partying, nor his supposed lack of interest in his studies, nor the strength of his religious convictions mattered. At Cambridge, Darwin took the first steps toward becoming a naturalist; in today's language, a practicing scientist. He began collecting beetles. In fact, he became so obsessed with this that, as Darwin recounted,

"one day, on tearing off some old bark, I saw two rare beetles and seized one in each hand; then I saw a third and new kind, which I could not bear to lose, so that I popped the one which I held in my right hand into my mouth. Alas it ejected some intensely acrid fluid, which burnt my tongue so that I was forced to spit the beetle out, which was lost, as well as the third one."

It was at Cambridge that Darwin realized what he wanted to do with his life. In his last year at the university, he read Personal Narrative of Travels to the Equinoctial Regions of the New Continent, During the Years 1799-1804 by the German naturalist and explorer Alexander von Humboldt (1769-1859) and A Preliminary Discourse on the Study of Natural Philosophy by the English astronomer Sir John Herschel (1792-1871). This *"stirred up in me a burning zeal to add even the most humble contribution to the noble structure of Natural Science,"* wrote Darwin. Darwin wanted to become a scientist. All he had to do was break the news to his father. Waiting for Darwin when he returned home from Wales was the letter that changed his life. In the letter dated 24 Aug, 1831 his friend and mentor John Stevens Henslow, had written,

"My dear Darwin,

Before I enter upon the immediate business of this letter, let us condole together upon the loss of our inestimable friend poor Ramsay of whose death you have undoubtedly heard long before this. I will not now dwell upon this painful subject as I shall hope to see you shortly fully expecting that you will eagerly catch at the offer which is likely to be made you of a trip to Terra del Fuego & home by the East Indies— I have been asked by Peacock who will read & forward this to you from London to recommend him a naturalist as companion to Capt Fitzroy employed by Government to survey the S. extremity of America— I have stated that I consider you to be the best qualified person I know of who is likely to undertake such a situation— I state this not on the supposition of yr. being a finished Naturalist, but as amply qualified for collecting, observing, & noting anything worthy to be noted in Natural History. Peacock has the appointment at his disposal & if he cannot find a man willing to take the office, the opportunity will probably be lost— Capt. F. wants a man (I understand) more as a companion than a mere collector & would not take any one however good a Naturalist who was not recommended to him likewise as a gentleman. Particulars of salary &c I know nothing. The Voyage is to last 2 yrs. & if you take plenty of Books with you, anything you please may be done— You will have ample opportunities at command— In short I suppose there never was a finer chance for a man of zeal & spirit. Capt F. is a young man. What I wish you to do is instantly to come to Town & consult with Peacock (at No. 7 Suffolk Street Pall Mall East or else at the University Club) & learn further particulars. Don't put on any modest doubts or fears about your disqualifications for I assure you I think you are the very man they are in search of—so conceive yourself to be tapped on the Shoulder by your Bum-Bailiff & affecte friend | J. S. Henslow

The expedn. is to sail on 25 Sept: (at earliest) so there is no time to be lost"

When the Beagle set sail on 27 December, according to one of Darwin's biographers, *"a new chapter in the history of science began"*. Darwin wrote in his autobiography that *"the voyage of the Beagle has been by far the most important event in my life, and determined my whole career"*. On 5 September, when Darwin had his father's permission and was soon to meet FitzRoy, he wrote to Henslow that *"Gloria in excelsis is the most moderate beginning I can think of"*. Darwin was excited about his upcoming adventure, and the research he did in the years between 1831 and 1836 established him in his career as a scientist and led to the writing of one of the most important books in the history of science.

Darwin wrote his thoughts on biodiversity and bio- geography to friends such as Henslow. His reflections and comments ranged so widely, from the formation of coral reefs to the ways in which seeds could be transported over the Pacific Ocean, that many naturalists were eagerly awaiting the publication of Darwin's findings from the voyage. Voyaging on the Beagle turned Darwin's life toward a career in science. Writing the papers, articles, and books based on his research during the five-year voyage made Darwin famous. And just for the record, Charles Darwin was already a well-known scientist fifteen years before The Origin of Species was published.

Darwin's life did not end after the publication of The Origin of Species. His name is so inextricably linked with the theory of evolution and the best-known exposition of that theory, The Origin of Species, that it is tempting to forget about Darwin after 1859. He did write The Descent of Man, but that book seems like a sequel to The Origin of Species which is easily overlooked. However, it is high time we stop using the term "theory" while mentioning Evolution. The term "theory" somehow makes some people think of Evolution as an unproven "hypothesis". Theory of Evolution is an incontrovertible fact of science. It is not a fictitious story like Creationism. It's a hard reality. It is the bed-rock of Biology. Defying evolution means defying one's own existence as a human being. Here, I'll not go into the life-long battle between Evolution and Creationism, that domain I leave on the strong shoulders of Richard Dawkins, one of the most interesting Evolutionary Biologists of our times. I can honestly say that he is perhaps the true and most capable advocate for Evolution. I'll just say one thing on this matter, anyone who doesn't accept evolution is perhaps still living in the stone-age. At least I'm proud to say that I and my cousin chimps come from the same ancestral line. And why not to be proud? After all, we fought really hard with the harsh and merciless environment for millions of years in order to achieve such advanced brain functions. Our present intricate humanly consciousness evolved after a long journey of struggle. And the beauty of natural selection is that our struggle against nature made us worthy of being rewarded with the 3 lbs. lump of highly advanced biological computer by our Mother Nature herself.

Meet Our Ancestors

The story of human evolution is the story of bones. Although it is a fascinating story, it is not a pleasant one. Most of our early ancestral bones emerged from the land of Africa. So, it's a very easy deduction that *"we are all Africans, because we all come from Africa"*. If everyone on this planet accepts this simple evolutionary fact, then the term "racism" would literally vanish from the face of earth.

We come from a long line of ancestors with different brain sizes. Some of them were smart, some not. There smartness was decided based on their brain size and circuits. But off-course none of them were as advanced as us. What do you expect! None of them had Wi-Fi, whatsapp, facebook or skype.

Human evolution began with the origin of life on Earth. The unicellular species represents the ancestors of life. However, around six million years ago our ancestral line diverged from the line which developed into modern chimpanzees and other apes. In this six million years period human brain evolved into what it is today. And the most obvious evolutionary change during human evolution has been the increase in the brain size. We'll discuss the evolution of the human brain in the next chapter. In this chapter we are going to meet our true ancestors.

In his book "The Descent of Man and Selection in Relation to Sex", Charles Darwin speculated that fossils of the earliest humans and their immediate progenitors ultimately would be found somewhere in Africa. He based his idea upon the fact that the natural range of our nearest living relatives, chimpanzees and gorillas, is limited to Africa. He concluded that we ultimately must have shared a now extinct common ancestor with those apes in Africa. Thereafter it was all about finding the missing link between apes and humans. And with each ancestral species we discovered, we came closer to discovering the exact ancestral species that'd ultimately join the two ancestral lines of apes and humans.

Darwin's book was published in the year 1871. However, like any new idea his thoughts were mostly rejected by the scientific community of that time. Before the 1920's, knowledge of our fossilized ancestors only went back to Homo Neanderthals or "the Neandarthal man" in Europe and some presumably earlier human-like forms from Java, in Southeast Asia. Only a handful of researchers were willing to estimate the time period of the earliest human ancestor at much more than 100,000 years, and there was no inkling of anything older from Africa. In addition, there was a bias among the predominantly European paleoanthropologists against accepting early Africans as the ancestors of all humanity, just like today's creationists.

Despite all the obstacles of the orthodox community the search for the human's ancestral species did go on far beyond 100,000 years ago and we started to discover our distant grandpas and grandmas. And finally we did trace back our ancestral line to Africa. Today the oldest ancestor we know of is Sahelanthropus tchadensis from Chad that lived between 6 and 7 million years ago (mya).

Now we'll look into each of the important species of our ancestral line. And we'll begin from the oldest known species of our ancestral line, i.e. Sahelanthropus tchadensis.

In order to understand the evolution of our species, we need to establish our ancestral state, i.e. what sort of animal we evolved from? For this purpose let's reconstruct the Last Common Ancestor of Humans and Chimpanzees. The Human-Chimpanzee Last Common Ancestor (HC-LCA) is the species from which the hominin lineage as well as chimpanzee & bonobo lineage diverged. However, it is not clear which, if any, of the fossilized species discovered so far represent the HC-LCA. That's why the search for the HC-LCA is still on.

Basically we come from Apes, which is the superfamily known as "Hominoidea". Hominoidea then branches into two families, which are the Hominids or Great Apes and the Gibbons or Lesser Apes. We humans belong to the hominid family along with other species such as orangutan, gorilla, bonobo and chimpanzee.

Among the great apes (chimpanzees, bonobos, gorillas, and orangutans), our closest relatives are the chimpanzees and bonobos. Fossil records, along with studies of human and ape DNA, clearly shows that humans shared a common ancestor with chimpanzees and bonobos sometime around 6 mya. We shall now have a closer look at our' evolution in Africa, from near the end of the geological time period known as the Miocene, just before our lineage diverged from that of chimpanzees and bonobos.

Hominins are the species on our branch of the hominid tree after the split with the chimpanzee & bonobo line, including us humans and all of the extinct species and evolutionary side branches. The first human-like traits to appear in the hominin fossil record are bipedal walking and smaller, blunt canines.

As I mentioned earlier the oldest species of our hominin sub-family is Sahelanthropus tchadensis from Chad. Sahelanthropus lived sometime between 7 and 6 mya, and had a combination of human-like and ape-like features. All that we know about this species is from only nine cranial specimens discovered in

Northern Chad by a research team of scientists led by French paleontologist Michael Brunet. In the year 2001 Brunet and his team unearthed the first and so far the only fossil specimens of Sahelanthropus. This partial skull was named "Toumai" ("hope of life" in the local Daza language of Chad in central Africa). Before 2001, early humans in Africa had only been found in the Great Rift Valley in East Africa and sites in South Africa. Naturally the discovery of this early hominin species in West-Central Africa showed that the earliest humans were more widely distributed than previously thought. Despite the lack of post-cranial bones (bones below the skull), the cranial fossil materials tell us enough to know that Sahelanthropus had both ape-like and human-like features. Ape-like features included a small brain of around 360 cc (even slightly smaller than a chimpanzee's), sloping face, massive brow ridges (similar in thickness to male gorillas), and elongated skull. However, the position and orientation of the foramen magnum, the hole in the base of the skull through which the spinal cord passes, suggests that Sahelanthropus stood and walked bipedally, with its spinal column held vertically as in modern humans rather than horizontally as in apes and other quadrupeds. Alongside the spinal cord opening underneath the skull, other prominent human-like features were small canine teeth and a short middle part of the face. They walked upright like humans which helped them survive in diverse habitats, including forests and grasslands.

After Sahelanthropus in the lineage of our hominin ancestors was Orrorin tugenensis which was discovered in the year 2001 from Eastern Africa. Orrorins were approximately the size of a chimpanzee and had small teeth with thick enamel, just like modern humans. In the Tugen Hills region of Central Kenya French paleontologist Brigitte Senut and French geologist Martin Pickford discovered more than a dozen early hominid fossils dating between about 6.2 million and 6.0 million years old.

Those fossils had previously unseen unique combination of ape-like and human-like characteristics for which they were given a new genus and species name, Orrorin tugenensis, which in the local language means "original man in the Tugen region". The upper femur of the species shows evidence of bone buildup that typically indicates one unique human-like quality, i.e. walking upright on two legs.

No skulls of Orrorin have been recovered, so its cranial morphology and brain size are uncertain. In both Orrorin and Sahelanthropus the canine teeth of males are larger and more pointed than in modern humans, but are small and blunt compared to the canines of male apes. This clearly suggests one thing, that competition among males for mating access to females was diminished in these early hominins compared to the great apes.

Sahelanthropus tchadensis and Orrorin tugenensis are undeniably the oldest hominin species known so far. But there is a lot to be learnt about them. So far the best known early hominins were Ardipithecus kadabba and Ardipithecus ramidus. Their lifetime dates back to right after Sahelantropus and Orrorin.

Ardipithecus kadabba was discovered in the year 1997 from Eastern Africa. Paleoanthropologist Yohannes Haile-Selassie didn't at first realize that he had uncovered a new species when he discovered a piece of lower jaw in the Middle Awash region of

Ethiopia. But the lower jaw specimen was followed by the discovery of 11 more specimens from at least 5 individuals, that paved the way for a new species. These specimens included hand, foot and toe bones, partial arm bones and a clavicle (collarbone). From the site of the fossil remnants faunal evidence was also found which implies that this early hominid ancestor of ours lived in a mixture of woodlands and grasslands. They had plenty of access to water via lakes and springs.

Originally those fossils were considered to be a subspecies of Ardipithecus ramidus. It was not until 2004, that those specimens from Ethiopia were allocated to a new species. Six fossilized teeth discovered in 2002 at the site Aso Koma made it confirmed that the 1997 specimens were indeed unique. Thereafter in the year 2004 paleoanthropologists Yohannes Haile-Selassie, Gen Suwa and Tim White published an article allocating those fossils to a new species – Ardipithecus kadabba where "kadabba" means "oldest ancestor" in Afar language. As far as bodily characteristics are concerned, they walked upright and were similar in body and brain size to a modern chimpanzee. They existed between about 5.8 and 5.2 million years ago.

Thereafter existed another species of the genus Ardipithecus about 4.4 million years ago. It is Ardipithecus ramidus that was discovered even before the discovery of the first kadabba fossil. In 1992–1993 a research team led by paleoanthropologist Tim White discovered the first Ardipitechus ramidus fossils that included seventeen fragments including skull, mandible, teeth and arm bones from the Afar region in the Middle Awash river valley of Ethiopia. More fragments were recovered in 1994, amounting to 45% of the total skeleton. At that time the most well established early human species were the Australopithecines. So those fossils were originally described as a species of Australopithecus, but White and his colleagues later in the year 2009 formally allocated them to a new genus, Ardipithecus as a new species Ardipithecus ramidus. In Afar language "Ardi" means "ground/floor" and "ramidus" means "root". However, the tern "pithecus" comes from the Greek word "pithekos" which means "monkey".

On October 1, 2009, paleontologists formally announced the discovery of the relatively complete A. ramidus fossil skeleton first unearthed in 1994. The fossil is the remains of a small-brained 50 kilogram (110 lb) female. And they nicknamed her as "Ardi".

Like most hominids, but unlike all previously recognized hominins, it had a grasping hallux or big toe adapted for locomotion in the trees. It is not confirmed how much other features of its skeleton reflect adaptation to bipedalism on the ground as well. Like later hominins, they had reduced canine teeth. A. ramidus had a small brain, measuring between 300 and 350 cc which is slightly smaller than a modern bonobo or female chimpanzee brain, but much smaller than the brain of australopithecines like the famous Lucy (around 500 cc) and roughly 20% the size of the modern Homo sapiens brain. Some of Ardi's teeth are still connected to her jawbone and show enamel wear that clearly indicates a diet consisting of fruit and nuts. And last but not the least A. ramidus lived around 4.4 million years ago.

Now comes one of the most important discovery of the paleoanthropological society.

It's the very much celebrated species of our beloved Lucy. A very important reason why Lucy is still so much celebrated is that she really embarrasses a lot of creationists. It really makes me pleased as a biologist when fundamentalists and creationists look up to Lucy as if she is not Lucy, but Lucifer the devil. And just for the record, there is no such thing as devil or satan or unicorn or sin. These are all constructs of the mesmerizing domain of the human brain. For further neuroscientific fun on this matter you can casually screw with my other two books The God Parasite: Revelation of Neuroscience and The Spirituality Engine.

Anyways, whether people believe or not, the harsh reality is that Lucy and Ardi are indeed our distant grandmothers. Now let's get acquainted with our fragile distant grandmother Lucy.

Her species is technically known as Australopithecus afarensis. They are one of the longest lived early hominin species. They lived between 3.85 and 2.95 million years. They lived like more than 900,000 years which is actually four times as long as our own species has been around. Lucy was discovered in 1974 near the village Hadar in Awash valley of the Afar region of Ethiopia by paleoanthropologist Donald Johanson. Lucy acquired her name from the Beatles' song "Lucy in the Sky with Diamonds", which was played loudly and repeatedly in the expedition camp all evening after the excavation team's first day of work. In the year 2007, Lucy's fossils were assembled and exhibited publicly in a six-year tour of the United States. This exhibition was called Lucy's Legacy. It attracted remarkable public attention and made Lucy a household name around the globe.

Au. Afarensis had both ape-like and human-like characteristics. They had ape-like face proportion (a flat nose, a strongly projecting lower jaw) with a cranial capacity of around 500 cc and long, strong arms with curved fingers adapted for climbing trees. And as for human-like characteristics, they walked upright on two legs and had small canine teeth. The fossilized specimens imply that Lucy's species was adapted for living both in trees and on the ground. Such adaptation proved amazingly beneficial for their survival throughout almost a million year despite the climate shift.

After Lucy there existed another early human species in the Australopithecus genus, known as Australopithecus africanus. Their timeline was about 3.3 to 2.1 million years ago. The first fossils of Au. Africanus were found at the Taung site near Kimberley, South Africa in the year 1924. They were the fossil remnants that made us embark on a completely unique early hominin genus, which we know as Australopithecus. But for such a discovery, the road ahead is always long and filled with hardship. And in this case, the hardship was about 20 years long.

In 1924, workers at the Buxton Limeworks near Taung, South Africa, showed a fossilized primate skull to E. G. Izod, the visiting director of the Northern Lime Company. The director gave it to his son, Pat Izod, who displayed it on the mantle over the fireplace. When Josephine Salmons, a friend of the Izod family, paid a visit to Pat's home, she noticed the primate skull, identified it as from an extinct monkey and realized its possible significance to her mentor, Raymond Dart. Josephine Salmons was the first female student of Dart who was an anatomist at the University of Witwatersrand.

In the hands of Salmons that skull found its way to Dart. And naturally it stuck him as unique, so he asked the company to send any more interesting fossilized skulls that should be unearthed. He examined the skull and completed a paper that allocated the fossil as a new species of a new genus - Australopithecus africanus which means "southern ape from Africa". He described it as "an extinct race of apes intermediate between living anthropoids and man". His paper appeared in the 7 February 1925 issue of Nature. Also, that young Au. africanus skull was nicknamed the Taung Child.

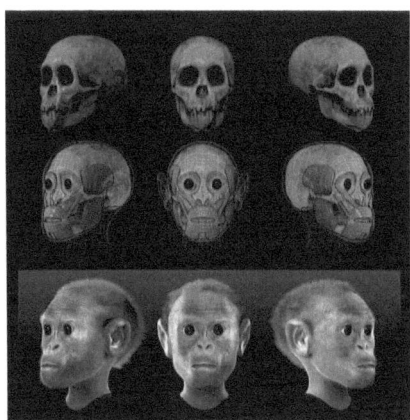

Figure 2-1: Taung child – Facial forensic reconstruction by Arc-Team, Antrocon NPO, Cicero Moraes, University of Padua

Dart's theory was supported by Robert Broom. On April 18, 1947, Broom and John T. Robinson discovered a skull belonging to a middle-aged female with a brain capacity of 485 cc, while blasting at Sterkfontein about 40 kilometers northwest of Johannesburg. Broom classified it primarily as Plesianthropus transvaalensis (near-man from the Transvaal) and it was nicknamed Mrs. Ples. It was later classified as Au. africanus.

Au. africanus was anatomically a lot similar to Au. afarensis. It had a combination of human-like and ape-like features. Compared to Lucy's species, Au. africanus had a more round cranium housing and smaller teeth, but it also had some ape-like features including relatively long arms and a strongly sloping face that projects out from underneath the braincase with a prominent jaw. Like Au. afarensis, the pelvis, femur and foot bones of Au. africanus clearly imply that they walked bipedally, but the shoulder and hand bones indicate they were also adapted for climbing. The fossil remnants show that they eventually left the trees for life on the ground except when chased back sporadically by the big cats which dominated the area. Because of their massive jaws, they were believed to have had a diet similar to modern chimpanzees, that consisted of fruits, plants, fibrous roots, nuts, seeds, fibrous roots, insects and eggs.

Even though no stone tools have been found in the sediments next to Au. africanus fossils, there are some indications in their fossils that they might have learnt to use stone tools. A 2015 study of hand bones in Au. africanus indicated that the species had "human-like trabecular bone pattern in the metacarpals consistent with forceful opposition of the thumb and fingers typically adopted during tool use," a pattern that would be consistent with an earlier adoption of tool manufacture and use than had been thought likely. However we need further studies on this matter to be absolutely sure that Au. africanus did indeed learn to manufacture first stone tools. However, in terms of first stone tool manufacturing, Homo habilis is called the "handy man".

Other species of the genus Australopithecus were Au. anamensis, Au. garhi and Au. sediba. We also find some evidence that Au. garhi learnt to make primordial tools as well.

Australopithecines were believed to be scavengers who ate fibrous roots, tubers, seeds, and vegetation. They received more useable calories out of the starchy tubers and vegetable foods than did tree-dwelling chimpanzees. Natural selection favored the genes responsible for the enzyme (amylase) for "grounded" hominids because this savanna diet was much more readily available than the ape's diet in the trees. Occasionally, some of these early hominids may have hunted small prey and broke open bones left by other animals with small pebbles from riverbeds. But they were definitely not efficient hunters even of small animals. Mostly they were foragers who fed off the leftovers from lions and larger cats. They also used bones for digging their roots and fibers. Tool use was not that different from contemporary chimpanzees. Some have estimated their average life span to be 30 years but children and females were particularly vulnerable to the many larger carnivores. It was still not at all a safe environment. Being upright meant that some of them could wield clubs for protection and carry food and other objects in their hands.

Eventually they left the forest altogether and moved to the savanna where their upright posture helped to see longer distances for scavenging food and watching for predators. Slowly hominid legs became longer and they developed arches in their feet allowing them to cover more ground than many of their four-legged cohabitants.

While living in the harsh environment of the wild, a very important tool of survival was being next to each other. This is what we call social organization. And our Australopithecine ancestors had a kind of basic social organization. The freaky surrounding compelled the Australopithecines to live in groups. Once they started to live in groups, they required further social skills in order to manage their social relationship, which in turn proved to be an important trigger for the increase in the brain size. By developing social skills the Australopithecines formed alliances and coalitions within the group in order to supervise their survival inside their society as well as outside of it.

With the limited cranial capacity of about 500 cc, the Australopithecines did indeed encounter the dawn of human consciousness. They were still very much primitive, at the same time they were primitively conscious. They developed a kind of emotional communication, which was confined to physical gestures and primitive vocalization. But, such communication system had its own headache. Any negative emotional outbreak could disrupt harmony in the group. Such emotional outbreaks were followed by a lot of noises which attracted attention of the predators. Naturally, as the theory of evolution suggests, this created an adaptive pressure for cortical control of emotion and for the so-called basic social emotions of sympathy, guilt, and shame which promote cohesiveness. This triggered an increase in the brain size which was mostly in the neocortex that added an extra layer to the whole brain and made room for more neurons. In fact their primitive form of social organization influenced the human brain to embark on an evolutionary journey of becoming the most social, emotional and advanced brain on planet Earth.

There is one commonality among all of us humans, which is, *we are all very much emotional.* The foundation of this emotion part goes way back to the time of our early Australopithecine ancestors. The deep-rooted instinct of emotions is founded upon that early sociality of Lucy and other Australopithecines. We are very much indebted to those primordial ancestors of ours, for gifting us this amazing social and emotional brain.

The first hominin species that developed the first sophisticated stone tools was the Homo habilis. They lived between 2.4 and 1.4 mya. In 1994 they were given the name "handy man". However, now we have evidence that older hominins like Au. garhi also developed primitive tools. Their tools may not have been as sophisticated as those of habilis but they were sufficient to break bones and expose the marrow which could remain eatable for a long time. This shift from a fruit enriched forest diet to a diet of scavenged meat set in motion a tripling of our ancestors' brains.

Sorry vegans. But if we had never started eating meat, we would've never become the smartest species on earth. Biologically speaking, meat is the best food for the brain. So, from an evolutionary perspective, meat diet was a great push towards evolution of modern human consciousness.

The main hardware that generates consciousness increased significantly in size during the lifetime of one of the early hominin species of our own genus "Homo", Homo habilis. During their lifetime of about 1 million year, their brain capacity increased tremendously from 550 cc to an astounding 800 cc. From a paleoanthropological perspective, Homo habilis was not that different physically from Australopithecine and could be better seen as a late Australopithecine.

Between 1960 and 1963 at Olduvai Gorge in Tanzania a team of paleoanthropologists led by Louis and Mary Leakey unearthed the fossil remnants of a unique early human species. The first specimen was found by Jonathan Leakey, so it was nicknamed "Johny's Child". As those early fossil remnants had a unique combination of features different from the Australopithecines, Louis Leakey, South African Scientist Philip Tobias and British Scientist John Napier declared them a new species, "Homo habilis", which means "handy man".

Similar to the Australopithecine forefathers, Homo habilis were scavengers and lived in social groups of around 70-80 individuals. For more social cohesion they required to develop further social intelligence. But they were still primitive and we still found no indication of language capacity. So communication was still limited to physical gestures, mimicry and primitive vocalization just like the Australopithecines.

Other than developing primitive stone tools, they didn't achieve any further technological or mental advancement. Like Au. garhi they survived on the dead animals' meat. They used their stone choppers to cut through the thick hides of dead animals and expose raw flesh. But the competition was huge. And those competitors were large and dangerous. Naturally, in order to win their meal, Homo habilis had to recruit new members in their group. This way the number of individuals in their groups eventually kept on increasing.

With an increasing number of individuals in their groups, their social skills slowly became more and more advanced. Their intense social interaction proved to be the foundation of the development of language in the brain circuits.

Now the human ancestral lineage had reached the crossroads of evolution. Among all our primitive ancestors, now arrived the one that really was a primitive representation of modern advanced humans. They were the Homo erectus. They solved one of the most important pieces of human evolutionary puzzle. It was the mystery of fire.

Every animal on earth that had ever encountered fire, had run away from it. If Homo erectus could do the unimaginable and conquer their instinctive fear they'd harness a new power. They just needed the nerve to reach into the blaze. The impact of fire was an enormous step forward in human evolution.

Despite such an advancement of fire, due to the lack of ample brainpower and vocal structure, Homo erectus were still prelinguistic. However, the brain structure did go through some remarkable changes during the whole period of their existence. Their cranial capacity doubled from 550 to 1100 cc. With the addition of further neocortex layers, the frontal, temporal and parietal lobes increased in size. Cognitive functioning was more focused on imitation and mimicry involving vocalizations, facial expressions, eye movements and above all emotional expressions.

They might not have been so advanced in terms of intelligence, but they developed highly effective emotional communication. The brain of Homo erectus was also lateralized to create two different hemispheres. At an intellectual level, they developed more refined, symmetrical and sharper tools. They made advanced weapons and tools like hand axes, cleavers, and knives. This gave them the capacity to free themselves from the dictatorship of the harsh environment and survive in the harsher climates to which they traveled. They had already learnt to make and harness the natural gift of fire. The remarkable technological advance of fire and refined tools gave Homo erectus heat, light and protection.

A Dutch surgeon named Eugène Dubois found the first fossilized Homo erectus in Indonesia in 1891. In 1894, Dubois named the species Pithecanthropus erectus that means "erect ape-man". At that time, Pithecanthropus (later changed to Homo) erectus was the most primitive of all known early human species. No early human fossils had ever been discovered in Africa yet. Homo erectus existed between 1.89 million and 143,000 years ago. It is possibly the longest lived early human species, about nine times as long as our own species has been around.

The most complete fossil individual of this species is known as the "Turkana Boy". Microscopic study of the teeth indicates that he grew up at a growth rate similar to that of a great ape. There is fossil evidence that this species cared for old and weak individuals. The appearance of Homo erectus in the fossil record is often associated with the earliest hand axes, the first major innovation in stone tool technology.

They are the oldest known early humans to have possessed modern human-like body proportions with relatively elongated legs and shorter arms compared to the size of the torso. These features clearly indicate adaptations to a life lived on the ground. They lost the earlier tree-climbing adaptations and developed the ability to walk and possibly run long distances.

Homo erectus is considered to have been the first species to have expanded beyond Africa. They migrated out of Africa to southern Asia and Europe about one million years ago. After developing advanced tools and learning to harness fire, they lived throughout the entire period of existence without any further advancement.

Over time, the body structure of our ancient ancestors became more robust with thicker bones and more muscle. And the perfect specimens for such robust body structure were Homo heidelbergensis and Homo neanderthalensis. At a point of time both these species competed with Homo erectus. The existence of heidelbergensis dates back to between 700,000 and 200,000 years ago, while the Neanderthals lived between about 400,000 and 40,000 year ago.

Homo heidelbergensis were the first early hominin species to routinely hunt large animals. In the year 1908 near Heidelberg, Germany in the Rosch sandpit just north of the village Mauer, the first fossil specimen was discovered. German scientist Otto

Schoentensack described the specimen as a completely new species.

This early ancestor of ours had a very large browridge, a larger braincase and a flatter face than all other previously known older hominins. Their robust body was the result of adaptation to colder climates. They took the torch of technological advancement one step further by building shelters for the first time in human history. As they migrated to colder climates, their bodies became more compact, which reduced overall skin surface area and heat loss. Such bodily adaption proved to be more efficient in conserving heat than a tall, lean body like Homo erectus, which exposed more surface area proportional to body mass useful in a hot, dry African environment.

Evidence suggests that H. heidelbergensis are the common ancestor of us Homo sapiens and our cousin species, the Neanderthals. Neanderthals diverged from H. heidelbergensis about 400,000 years ago in Europe and lived until about 40,000 years ago, while we the sapiens evolved about 200,000 years ago in Africa during a time of dramatic climate change.

As our cousins, Neanderthals are our closest extinct relatives. However, among the living relatives the closest are the chimpanzees with 96% genetic similarity. But the common ancestor of ours and the chimpanzees goes way back millions of years, while the common ancestor of ours and the Neanderthals existed as close as 200,000 years ago.

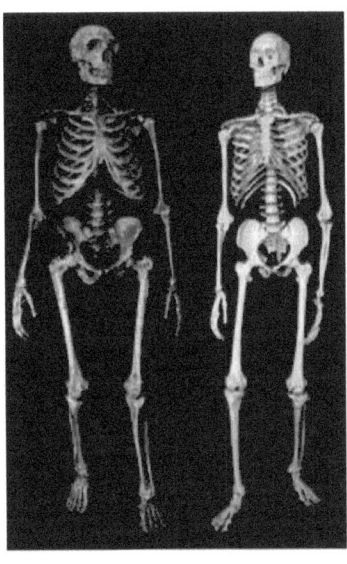

Figure 2-2: Left: Neanderthals, Right: Homo sapiens (us)

The Neanderthals had a larger headcase than us the sapiens. Their cranial capacity was an average of around 1500 cc, which is much larger than the average 1300 cc of modern humans. In the year 1829 the first specimens of fossilized Neanderthal skulls were discovered in Engis Caves, Belgium.

Then again in 1848 another skull was found in Forbes' Quarry, Gibraltar. But it was not until 1856, that those fossils were recognized as a new species. Johan Karl Fuhrott first recognized those fossils as "Neanderthal man", after the discovery of another similar kind of fossil in Neander valley near Mettmann, North Rhine-Westphalia, Germany. After this discovery, Geologist William King proposed the name "Homo neanderthalensis" at a meeting of the British Association for the Advancement of Science in 1864.

The tall, lean bodies of Homo erectus were adapted to tropical temperatures. Whereas late hominin species Neanderthals were adapted to winter climates. Neanderthals were more heavily built than modern humans. With their huge cranial capacity of around 1500 cc they developed extraordinary taste in aesthetics. They occasionally made symbolic or ornamental objects. A very unique characteristic of Neanderthals was their primitive form of spirituality.

Paleoanthropological evidence clearly shows that they deliberately buried their dead and even marked the graves with offerings, such as flowers. This implies that our cousin Neanderthals believed in some form of life after death. No other primates, and no earlier human species, had ever practiced this sophisticated and symbolic behavior. They may not have been so advanced as us, but among all the extinct hominin species Neanderthals were exceptionally advanced. They were amazingly rich in behavior. In terms of modern human consciousness, Homo neanderthalensis were right next to us.

Neanderthals were capable of symbolic thinking. Archeologists have unearthed shells containing pigment residues at two sites in the Murcia province of Southern Spain. Black sticks of the pigment manganese, which may have been used as body paint by Neanderthals, have previously been discovered in Africa. The pigment containers are the first significant evidence for their use of cosmetics. Until this discovery in the year 2010, it was thought that only the humans were conscious about their looks. But now it seems that Neanderthals really did engage in grooming their looks as well.

Their tools were unlike anything else made by other extinct hominins. These tools consisted of sophisticated stone-flakes, various hand axes and spears. They were even expert in building dugout boats, in which they navigated the Mediterranean Sea. They were pro hunters, who obtained most of their protein in their diet from animal sources. They knew how to create and control fire. They used caves for shelters as well as built beautiful homes using animal bones. And of course with their sophisticated tools they also made clothing. It has been speculated that Neanderthals had a proto-linguistic system of communication that was more musical than modern human language.

We humans existed side-by-side with Neanderthals for some time. The very first thing you must know about Homo sapiens is that the whole humanity emerged in Africa about 200,000 years ago. So, to anyone who knows the true history of mankind, "racism" is just an insignificant piece of social junk. According to genetic and anthropological evidence Homo sapiens evolved exclusively in Africa. So, we are all originally Africans.

You are human yourself, so I guess I don't have to elaborate on the amazing cognitive abilities of Homo sapiens. You already know all about the powers of the human mind. The binomial name Homo sapiens was coined by Carl Linnaeus in the year 1758. The Latin noun "homo" means "human being". So, regardless of all our inhuman behaviors in our modern society we'll still be called "human beings" in the name of science.

Sp. Note:

Figure 2-3: Homo naledi (Reconstruction by John Gurche)

The tree of human family has a lot of species. Among them I have described the most important ones that went through significant evolutionary developments. We still don't know all about our early hominin ancestors. But we keep learning and often dig up new species with unique structural characteristics. The most recent discovery we have made is an early hominin species named Homo naledi.

The first naledi specimens were discovered in 2013 at the Dinaledi Chamber of the Rising Star Cave system, part of the Cradle of Humankind World Heritage Site, South Africa. In September, 2015 paleoanthropologist Lee Berger allocated all the 1550 fossil specimens discovered until that month to a new species "Homo naledi". Even though these fossil specimens have not yet been dated, judging from the timing and characteristics of other specimens of early Homos and late Australopithecines, I deduce that Homo naledi existed sometime around 2-3 million years ago.

Thus, we keep getting acquainted with our true ancestors. So it is imperative that we keep digging, literally.

Evolution of Human Brain

"I know, my dear Watson, that you share my love of all that is bizarre and outside the conventions and humdrum routines of everyday life."

- Sherlock Holmes

I'm proud to say that we humans are the only species on earth that has ever asked the question,

"Where do I come from?"

And such curiosity is only possible because of a uniquely special human consciousness. And the credit for the creation of human consciousness goes to the *Human Brain*. It is the organ of unparalleled importance. It is the organ of creation. It is the organ with which you see, perceive, observe, think, recollect, reminiscence, pleasure, love, kiss, make love, make babies, see God, feel God, create God, create a nation, create a world and even create an artificial version of yourself. In short, Human Brain is the only God on planet earth. Technically we are the brain.

Essentially, the evolution of the human brain has been one of the most significant events in the evolution of hominin life. It has been a 6 million years long mosaic process of size increase laced with episodes of reorganization of the cerebral cortex. And the human brain is *"the most intricate, complicated and impressive organ ever to have evolved"*. The most obvious evolutionary change during human evolution has been the increase in size and complexity of the human brain.

The study of the evolution of human brain is an intriguing domain of scientific exploration. It's like talking to the ghosts of our extinct ancestors through their fossil remnants. This is what we call "Paleoneurology". Paleoneurology allows us to look into the details of brain structure of extinct species through close observation of endocasts (endocranial casts). It is a subfield of paleoanthropology.

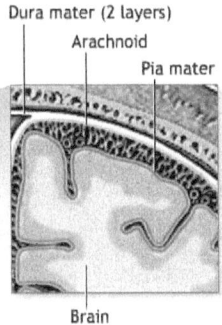

Figure 3-1: Three Meningeal Layers surrounding the Brain

Paleoneurological study is at present the only direct line of evidence through time. The main object of study here is the endocast that is simply cast made from the inside of the cranial bone. But there is a very important factor to remember. It is that endocasts are not exactly casts of the brains. When alive, the brain is surrounded by three meningeal layers, the dura matter, arachnoid tissue and lastly the pia matter, a thin layer of tissue directly overlying the brain. Upon death, all these layers of tissues as well as the brain dissolve and leave a cranium that eventually fossilizes in time. However, technically endocasts are so far the only direct way of exploring the evolution of the human brain.

For over a century, paleoneurologists have focused on intricate analyses of endocasts from human ancestors. Today the evolution of neurological reorganization is assessed in light of findings from paleoneurology. We can even make inferences about cognitive evolution through the interpretation of paleoneurological data within a framework that incorporates behavioral information from comparative primatological studies and findings from comparative neuroanatomical and medical imaging investigations.

Endocasts reproduce a good deal of information about the brain, including its general shape and details of some of its associated blood vessels, cranial nerves and cranial sutures. Endocasts even reproduce information about the convolutions of the brain that were imprinted on the inner walls of the braincase during life.

The convolutions of gray matter on the brain's surface, consist of bulges (gyri) and the grooves (sulci) that separate them.

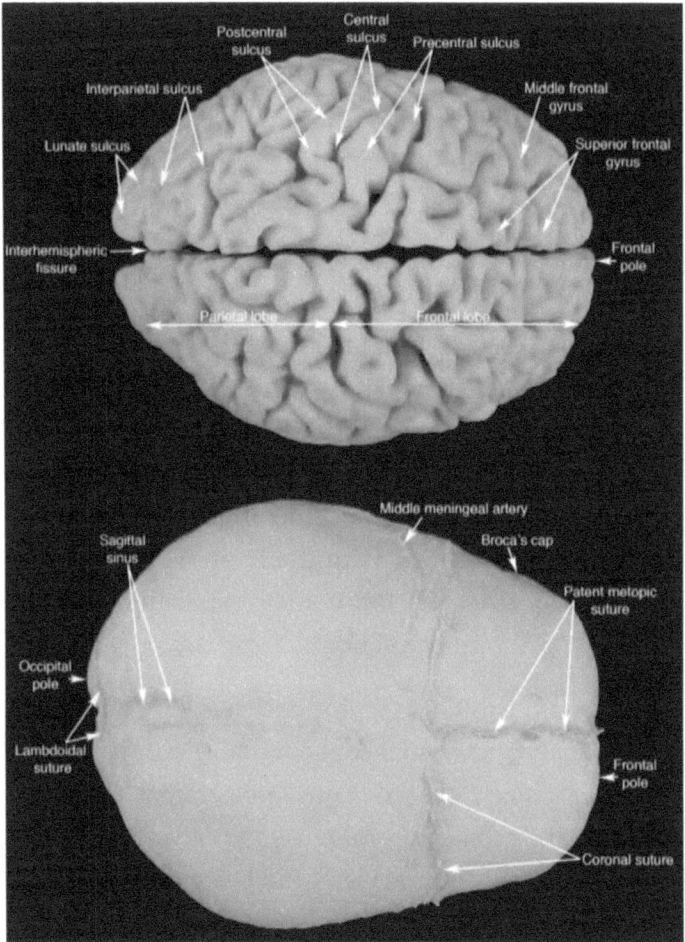

Figure 3-2: A dorsal view of a cast of a modern human brain and its accompanying endocast

The most useful data we receive by studying endocasts is the size of once living brain. The brain volume is usually determined by either water displacement of the endocasts or by a computer algorithm which simply adds sections taken from a CT scan of either the endocasts or the cranium.

Endocast volumes are somewhat larger, by about 8–12%, than the actual once-living brain, as the endocranial volume (ECV) includes meninges, cerebral fluid and cranial nerves. Hence, cranial capacities should be decreased by a corrective factor to compensate for the volume of the fluids and meninges that occupy the braincase along with the brain. However, it is quite common for cranial capacities to be used without correction as proxies for brain size.

By analyzing cranial capacities and estimates of body size (based on postcranial fossil remnants), paleoneurologists hypothesized that both the absolute mass of the brain and its size relative to body mass (relative brain size, RBS) increased independently during the evolution of the primates, as well as other mammals. Through the analysis of the fossil record, in concert with a comparative neuroanatomical analysis of closely related species, it shows that the hominid brain increased in size more than threefold over a period of approximately 2.5 million years. However, it has become increasingly clear that the human brain is not simply a large ape brain but this 3 lbs. cauliflower also developed important qualitative and quantitative changes.

Among all the hominid species, Neanderthals had the largest brain, which was an average of about 1500 cc. We Homo sapiens now have a brain size of an average 1300 cc which is nearly three times the brain of our dearly beloved distant grandmother Lucy. Recent quantitative analysis of the endocasts has verified that naturally selective pressures for enlarged brains began early in primate evolution but has also revealed that brain size decreased independently in some branches of old monkeys and new world monkeys.

As you can see, our cousin species Neanderthals had larger brain than us Homo sapiens, yet they couldn't become the smartest species on earth and got extinct. Also, in terms of size among other mammals the 1.35 kg brain of Homo sapiens is significantly exceeded by the brains of elephants and whales. This implies that a larger brain does not necessarily assure greater consciousness. Rather, in order to develop the unique software of human consciousness, the brain went through evolutionarily unique transformations in terms of size, neural organization and proportions of different brain regions. All these factors along with the help of the right kind of niche influenced the evolution of consciousness.

In order to properly understand the properties of the brain among various species, numerous analytical techniques and measures have been developed, such as Absolute Brain Size, Relative Brain Size (RBS) and Encephalization Quotient (EQ) etc.

Among the living primate brains, the human brain is the largest, weighing about three times more than the brains of our closest living primate cousin, the chimpanzees. This is what we call an absolute difference in brain size. The total weight or volume of the brain, i.e. the Absolute Brain Size can tell us scientists a great deal about the attributes of a particular species. Absolute Brain Size has been shown to predict cognitive and behavioral flexibility in nonhuman primates. The more flexible a species is, the higher their intelligence. And the higher the intelligence is, the more advanced and unique the consciousness. Just as brain size increases from prosimians to monkeys to apes, so does the cognitive and behavioral flexibility of these animals.

There are certain minimum requirements for brain size in all mammals. At the lowest level, there must be enough brainpower to keep the body in homeostasis that is to maintain internal equilibrium even during the environmental changes. There must be enough brain tissue to sense the outside world and to respond appropriately to the stimuli from both outside and inside. These basic functions allow a species to feed, flee danger, and reproduce. Animals that have more brain tissue than just the basic level necessary for survival have an advantage over animals that survive with the minimum requirements for their body size. In order to determine which animals have more brainpower than is strictly necessary, we must know the relationship between the size of the brain and the size of the body. This is what we call Relative Brain Size (RBS). There are many ways to define RBS but the simplest is the weight of the brain divided by body weight. Modern humans have an RBS of roughly 2%. Among large mammals, humans have the relatively largest brain, whereas shrews, the smallest mammals who exhibit supposedly much less cognitive and behavioral flexibility have brains up to 10% of their body mass.

Small-bodied animals like mice have relatively large brains for their body size, while large-bodied animals like elephants or whales have relatively small brains for their body size. We see a similar phenomenon during growth. A small-bodied baby has a much larger head in relation to its body than a large-bodied adult does. This same relationship holds true for primates. As primate body size increases, the ratio of brain size to body size decreases.

When body and brain sizes of different animals are known, we can calculate an amazing measuring statistic called the Encephalization Quotient (EQ). But what is EQ exactly? For example, if we draw a relationship between body weight and brain weight from various primate species, we get to see something beautiful. Let's consider the brain and body weight of lemurs on this matter. They have much brain than would be predicted for a primate of equal body size. And if we consider ourselves, we find out that we have much larger brain than would be predicted for a primate of equal body size. From this we can deduce that, humans are much more encephalized than the lemurs. In fact, humans are more encephalized than all other primates.

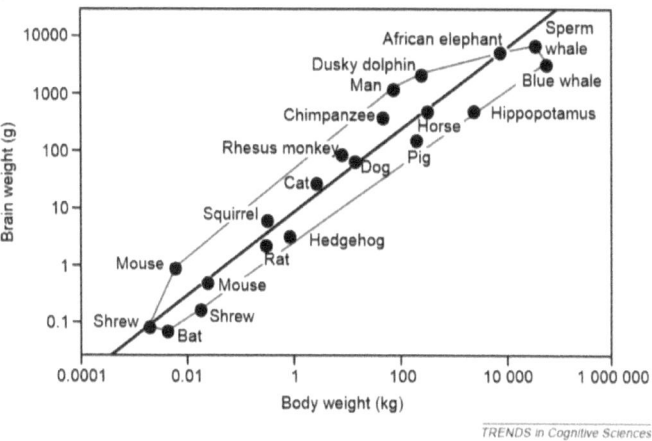

Figure 3-3: Relationship between brain size and body size in selected mammals. Brain size (kg) and body size (g) are given for 20 mammals, including those with the smallest and the largest body and brain weights (in the shrew and mouse, two different species are represented). In all vertebrates, brain size increases negatively allometrically with a power function of exponent 0.6–0.8, meaning that an increase in brain size lags behind an increase in body size.

Harry Jerison, one of the world's foremost experts in brain evolution and intelligence, first developed Encephalization Quotient to compare how much a species diverges from the line of predicted brain size. Jerison's line included data on a large sample of living mammals. The EQ is thus simply the ratio of actual brain size to predicted brain size for Jerison's mammalian data set.

EQ can give us a basic idea on how much more or less brainpower a species has in comparison with other mammals. And naturally, among all the mammalian species primates have high EQ with humans on top of all.

Feel proud my friend, you are scientifically the smartest species on earth.

Brainpower is highly dependent upon the diet. And the science of EQ noticeably depicts that only a vegetarian diet doesn't ever smoothen the development of cognitive capacity of a species. For example, monkeys that eat leaves have lower EQ than monkeys that eat fruits or apes that are omnivorous. And fruit-eating monkeys and omnivorous apes have lower cognitive capacity than primates that eat abundantly. Late Australopithecines and early Homos had a varied diet that included all sorts of food from meat to fruits and tubers. Such a vivid nutritious diet literally impacted greatly over the encephalization of the human brain. As a result, today's highly advanced human brainpower evolved after a significant evolutionary transformation over a period of about 2.5 million years. With this amazing brainpower humans do way more than just survive.

With the limited brainpower the Australopithecines and early Homos like Homo habilis did indeed have a primitive form of consciousness, which enabled them to respond to the environmental changes. Some of them even made primordial advancements in stone technology. Over time as the human brain went through a mosaic process of reorganization of brain tissue and the entire brain circuitry, it became more and more complex. This ever-increasing complexity of brain wiring called for the dawn of modern Human Consciousness.

Evolution of Speech

In terms of thinking, do you know what is most unique about the human species?

The answer is *"Imagination"*.

The moment our brain developed the biological capacity of imagination, all of a sudden a completely new universe opened up to us. And such imagination was possible by the arrival of language and speech. Rather than living only in the present, the use of verbal symbols allowed the Homo sapiens to mysteriously transcend the immediate experience given by the physical senses and to live in an abstract, extra-sensory, and hypothetical world. And when the brain was able to generate virtual hypothetical situations in its circuits, it was a great leap forward in the path of attaining modern human consciousness. They had to develop cognitive abilities to transcend immediate sensory experience, because it was necessary to free themselves from the dictates of the environment.

We know now that humans are capable of amazing abstract thinking, but how exactly can we know when it all began? For this, we again turn towards fossil records. Homo sapiens brain reached its present average size of 1300 cc fully expanding the frontal lobes making symbolic thoughts possible. And the larger left hemisphere allowed more space for the regions required for speech production, namely Broca's area (in the frontal lobe) which houses the capacity to produce grammatical speech and Wernicke's area (behind the temporal lobes) which makes possible the semantic understanding of words and the reception of speech.

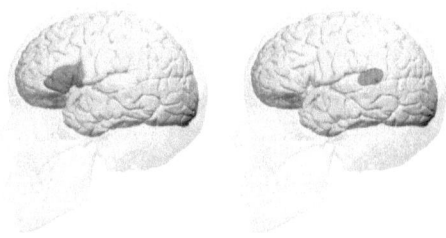

Figure 4-1: Left: Broca's Area, Right: Wernicke's Area

Paleoneurological evidence suggests that only in Homo sapiens did the part of the brain crucial to language become vascularized enough to allow the needed blood supplies. This was especially true in Wernicke's area. This process for a long time had been developing in earlier hominids but had not reached the critical stage where it could provide a base for language development. After detail analysis of the fossilized skulls paleoneurologists traced the distribution of blood supplies for the lateral surfaces of the skull where indentations were made by the blood vessels. Such vascularization in both hemispheres of the brain were exactly what the humans needed to develop various cognitive abilities. In general the left hemisphere is associated with sequential integration of experiences and is strongly intertwined with language processes. The sense of self is strongly correlated with language and consequently with the left hemisphere. The right hemisphere is structurally organized for more simultaneous integration of various experiences. While the left hemisphere produces your sense of self, the right hemisphere on the other hand functionally organizes information as a function of vigilance and anticipation. Technically, the right side of your brain enables you to perceive "the big picture". Ergo, the right hemispheric vascularization triggered significant increase in aesthetic sensitivity. The left hemispheric cognitive abilities and the aesthetic abilities of the right hemisphere together made a species distinctively human.

Social organization was imperative for the early hominins to survive in the harsh environment. So, natural selection forced the brain to develop primitive social interaction through gesture and mimicry. And thereafter the brain developed intelligence. Such selective natural pressure gave the Australopithecines cortical capacity for social coordination. Any negative emotional outbreak would attract the attention of the predators, so our early ancestors had to develop emotional control. As the early humans gained control over their emotions by the augmentation of the prefrontal cortex, their social communities became more stable. With their early brain capacity of very primitive social communication the late Australopithecines built tools and later the Homos built shelters.

Such technological advancement precedes the arrival of an efficient language. This implies that early humans developed pretty interesting primitive form of consciousness. But the threats of the environment demanded those early humans to become more advanced by developing further logical thinking. There is plenty of evidence that logical thinking occurred before language. All the environmental demands triggered significant enlargement of the prefrontal and temporal cortices. The prefrontal cortex enabled instrumental behavior, concentration, and emotional control as well as the integration of cognition and emotion necessary for decision-making and analytical thinking. Throughout a period of 6 million years there was a remarkable increase in cerebral cortical connections involved in cognition. Augmentation in the neocortex ultimately facilitated amazing brain functions to produce experiences of a coherent and meaningful world.

Early hominins such as Australopithecines and early Homos all had gesture oriented and emotional communication among each other.

But it was the Homo sapiens that had the biological readiness for language as early as 150,000 years ago, but readiness does not suffice to produce speech. Most commonly cited date for the so-called origin of language is about 70,000-35,000 years ago. But all the Homo sapiens did not develop speech at that time. The origin of speech was depended on a beautiful amalgamation of cultural, symbolic and anatomical readiness.

The expanded cortical layers of the temporal lobes house Broca's area which is associated with the production of speech in the frontal lobe and Wernicke's area further behind the temporal lobe allowing the understanding of speech. Another evolutionary anatomical change necessary for the production of speech is the capacity of the throat and tongue to make a vast number of sounds. This capacity depends on the "descent of the larynx" which is necessary to facilitate these speech movements and possibly have started with the genus Homo. The key to this descent is the very small hyoid bone within the larynx. Because the hyoid is so small the fossil evidence for it is scarce. However, paleoanthropological evidence shows that the Neanderthals had the hyoid bone as well. Yet due to the lack of proper readiness of the brain's language and speech centers, the Neanderthals were still proto-linguistic, which means the system of their communication had some but not all elements of language. But eventually the Homo sapiens developed all kinds of perfect biological readiness to be adorned with the capacity of sophisticated speech and language, for the first time in hominid history. Hence we humans became truly unique.

Nature of Consciousness

What is Consciousness?

The simplest answer to this question is – Consciousness is the functional expression of the brain.

However, we have only started to scratch the surface on this matter. Science has finally started to understand the true nature of consciousness. However, if someone says he understands Consciousness fully, he doesn't know jack about Consciousness. Consciousness is a topic on which a lot has been written and discussed, but there is still a lot to investigate and learn.

Till this date consciousness remains an enigmatic mystery to the general population. Nevertheless, with the advancement of neuroscience in the last few decades, we have started our journey to explore the true biological foundation of Consciousness. And even though, the mystics and some of the hardcore philosophers want you to believe otherwise, the foundation of consciousness lies in the complex interplay of the brain circuits. The air of mystery surrounding consciousness has provided the mystics, means of survival for ages, but as we neuroscientists keep on exploring the mysterious domain of consciousness the majority of mystery-mongering professionals feel threatened. Because the very existence of them is predicated on the metaphysical approach towards explaining consciousness. But as the journey of neuroscientific exploration goes on, one day we shall know all there is to know about consciousness without the need of metaphysical explanations.

Despite the lack of ultimate understanding of consciousness it feels good to be the owner of it. Above all, it's really special to be humans endowed with the evolutionary gift of Consciousness. But what is so special about humans?

We are special because, we are never satisfied with anything. As my friend Mathematician Ronald Cicurel says, *"we are always curious"*. Now the question that rises is "where does curiosity come from?"

The fuel for the evolution of curiosity was ignorance. To pacify ignorance our primitive ancestors developed curiosity and many other evolutionary characteristics. Among all the evolutionary characteristics, the most crucial ones are Curiosity, Spirituality and Sexuality. All these traits serve a single biological purpose, i.e. *"survival of the species"*.

All these traits hand in hand construct the human consciousness. Consciousness is simply about everything we are. Every living being on planet earth has some form of consciousness. It's the most common quality of life. Among all other species, humans are the only one who have crossed the limitations of primitive consciousness and have become the smartest of all. Even though, each person's consciousness is uniquely different from another, there is something similar about it in everybody, that's why we understand each other.

We are all separately unique at our conscious, subconscious and unconscious levels. And all these levels of the human mind are born in the human brain. On a cursory observation of the neural networks, it may seem to be similar in all of us humans, but if we go deeper and observe every subtle detail of the neural wirings, we shall find that each human is quite uniquely wired. That's why some individuals develop passion for science, some for music and some for philosophy etc.

A common misperception among people about the human brain is that we are hardwired to do this and that. But the reality is there is nothing fixed in the brain. There is no hardwiring in the brain. Or as my fellow neuroscientist David Eagleman says, *"we are not hardwired, we are live-wired"*.

We are constantly changing. For example, by the time you have read this page, your brain circuits have been rewired, since the moment you started reading this book. Human consciousness is constantly receiving information from the surroundings. Every single second, the human mind keeps evolving. Even while you are sitting in your comfy couch with this book in your hand, your brain is still changing. No matter what you do or do not do, the brain never stops rewiring itself. It's like every single second you are a new person. If a moment ago you were smart, now you are even smarter.

And no matter what kinda fortune cookie this "Consciousness" thing is, it is still all in your brain. But, since your consciousness has evolved over time, you are not the same you with whom I started talking at the beginning of this book. Each new second can be an opportunity for salvation, redemption, reformation or enlightenment. Whatever you do, you do it with your consciousness, which is a constantly evolving product of your evolving brain.

We have only started to scratch the surface in the path of understanding the human brain and thereafter human consciousness. So, the so called concepts of "Singularity" and "Transhumanism" are nothing but a fairy tale. At this point claiming that we can simulate the human brain functions without the correlated human biology, is like the Homo naledi saying *"we'll soon invent an actual Time Machine"*.

Transferring human consciousness into a digital system is only a tall tale. Nothing can live on computers. Computers can record and calculate, but only the human brain can create. Till this date, a computer does not understand what it does. Computer scientists can program a computer to be so smart that it can actually act as if it is conscious, but it'd still be a programed behavior of pretense. It is good to have an open mind but not so open that your brain falls off.

"Human mind may have invented the computer, but it has not invented the computer that can invent a human mind, nor will it ever be able to do so."

- Sir Roger Penrose

Singularity is nothing but a concept of technological industrialization over the human mind. Saying that human consciousness can live on computers actually makes people think more mechanically than ever and they literally become psychologically dependent over technology. This gives the technological industry means to make more money just by manipulating people's sophisticated desires.

However, this chapter is not about describing the sophisticated fallacies of Singularity and Transhumanism. Rather this is about the fascinating nature of consciousness. So, let's get back to the topic.

If we observe close enough we can find that every living organism on planet earth has some sort of consciousness about their existence. At the very least, this primordial form of consciousness allows a species to react to environmental changes in order to ensure survival. In this context, I remember a fascinating story of an experiment carried out by an Indian scientist, that is still taught in the elementary schools of Bengal. His name was Jagadish Chandra Bose. I just thought of mentioning him, because this experiment of his, concerns a basic form of consciousness in plants.

He was mainly a Biologist whose major contribution in the field of biophysics was the demonstration of the electrical nature of various stimuli (chemical agents, wounds etc.) in plants, or in simple terms, he showed that plants react like humans to various physical stimuli like pain, affection etc. Bose's perception was that all plants are endowed with a certain degree of individuality. By means of the Resonant Recorder and the Electric Probe designed by him, Bose was able to demonstrate that the collapse in the leaves of the Mimosa plant upon stimulation was accompanied by an electrical signal which traveled to the stem, and its passage through the stem (in both up and down directions) caused the other leaves to collapse. Similar experiments confirmed the coupling of electrical oscillations and spontaneous leaf movements in Desmodium.

These experiments led Bose to the conclusion that just like animals, plants are in possession of a nervous system. The response of the whole plant to physical stimuli is a consequence of long range electrical signaling throughout the entire plant body. This is similar to the basic characteristic of neurobiology in animals.

So, we can scientifically confirm that plants have a primordial form of consciousness. However, such plant consciousness is termed as "Tropism". Tropism refers to directed response of a plant due to environmental stimuli. There are various types of Tropism

Phototropism - Response to light

Heliotropism – Diurnal or seasonal response to the direction of the sun

Gravitotropism – Response to gravity

Thigmotropism – Response to touch

Hydrotropism – Response to water

Chemotropism – Response to chemical agents

Thermotropism – Response to temperature

Electrotropism – Response to an electric field

There is also a weird kind of tropism in the characteristic of the parasitic creeper vine Monstera which grows along the ground in search of darkness caused by the shadow of a prospective host tree. This is called Skototropism. It is something like negative phototropism.

Similar to the neurobiology of animal anatomy, Tropism involves three distinct phases

1. Detection of the initial environmental signal by plant receptors,
2. Subsequent processing/transduction of the primary signal,

3. The consequent integrated physiological response.

Consciousness in plants manifests through electrochemical signals, just like in humans. There are actually two forms of electrical signals prevalent in plants: the action potentials (AP) and the slow wave variational potentials (VP). In contrast to the action potentials, the VP's vary with the intensity of the stimulus and have delayed repolarizations. The ionic mechanism behind the transmission of AP's also differ significantly from VP's. Both AP's and VP's are involved in long distance signaling and can invoke a response distant from the local area of the applied stimulus. AP's have been implicated in very rudimentary forms of conscious activities such as trap/tentacle closure (for Dionaea, Drosera), regulation of leaf movements (Mimosa), increase in respiration and gas exchange (Zea), decrease in stem growth (Luffa) and induction of gene expression (Lycopersicon).

However, as plants are rooted to the same spot all through lifetime, the physiological response is more in terms of growth and development in contrast to animals who respond primarily by a variety of movements.

With the increasing complexity of the nervous system, a species climbs up the ladder to become more advanced in terms of consciousness. So far, the only species that has reached the top of the ladder is us, the Homo sapiens. Scientifically speaking, even the plants are conscious and so is the entire animal kingdom. But you never get to hear that a chimpanzee has written a book on consciousness. While surfing on the net you never come across some extraordinary news like, Simba – The Lion King has published a paper on the evolution of lion consciousness. It is only us who can perform such amazing feats of intelligence. And it is all by the grace of the cute little human brain inside your skull.

All our thoughts, feelings, emotions, ideas, even the sense of our intimate self emerge exclusively from the intricately beautiful activity of the human brain. And this very fact often embarrasses the sophisticated human vanity of some philosophers, such as Slavoj Zizek and many more.

It's not just the mystics and philosophers, rather most of the human society finds it difficult to reconcile a picture of a physical matter that makes up the brain with a vivid consciousness that appears to result from its action. And surely they think, that mind and body are totally distinct things. What link can there be between that spongy grey matter of cells and the glorious human consciousness? As we brain scientists keep on exploring various mystical realms of the biological basis of human consciousness, we find it impossible to separate the subjectivity of the human mind. After all, we don't have any private nonmaterial objective conscious perception of the reality whatsoever.

Once you realize that far from being a spectator, you are simply vibrating with the eternal ebb and flow of cosmological events, you'd attain true absolution.

Qualia & Human Reality

"Four blind men went to see an elephant. One touched the leg of the elephant, and said, 'The elephant is like a pillar.' The second touched the trunk, and said, 'The elephant is like a thick stick or club.' The third touched the belly, and said, 'The elephant is like a big jar.' The fourth touched the ears, and said, 'The elephant is like a winnowing basket.' Thus they began to dispute amongst themselves as to the figure of the elephant. A passer-by seeing them thus quarrelling, said, 'What is it that you are disputing about?' They told him everything, and asked him to arbitrate. That man said, 'None of you has seen the elephant. The elephant is not like a pillar, its legs are like pillars. It is not like a big water-vessel, its belly is like a water-vessel. It is not like a winnowing basket, its ears are like winnowing baskets. It is not like a thick stick or club, but its proboscis is like that. The elephant is the combination of all these.'"

This story I heard from my teacher, or rather the only true teacher that I had. And it is quite relevant in the context of this chapter.

The construct of reality by the human consciousness is a combination of fascinating neurological processes inside the brain. Therefore whenever the brain circuits malfunction due to various neurological conditions such as temporal lobe epilepsy, synesthesia, anosognosia, Cotard's syndrome, Capgras' delusion, etc., the consciousness tends to malfunction as well. In turn that defective consciousness leads to an apparently defective reality. In this chapter through a fascinating analysis of various neurological conditions we shall scientifically deduce that the perception of reality indeed solely arises from the activity of a small subset of neural circuits, as suggested by Francis Crick - *"we are all pack of neurons"*.

So, in simple terms, what you perceive as real, is actually a neurological reconstruction or simulation of the actual real thing. It's not as simple as saying, we see as it is. Actually we do not ever see as it is. Reality is not recorded in our brain. We humans construct our own Reality based on our own presuppositions, hypotheses, beliefs and limited knowledge.

In fact, the true reality is mostly unknowable and will always stay that way. Because everything we perceive is always subjective. And even if we find a way to observe a material event objectively, we may never be able to understand the true nature of that event. All that we receive from being objective is data. And most importantly such data is always human interpreted. So, the true essence of the reality is lost in translation and interpretation.

To be objective, we need to form measuring scales. For example to study the reality of the universe physicists formed the measures called "Time" and "Space". To be honest, as a species we may not live long enough to ever understand the true reality of the universe. But as the foundation of consciousness lies within the domain of biology one day we shall know all there is to know about it.

Science always works on hypothesis. A scientist first has to presuppose any theory based on available relevant data. And you can realize how daring such presupposition can be, if you look at the history of Quantum Physics. In terms of such presupposition, a scientist begins his journey of scientific exploration as a philosopher.

And even our field of biology is founded on another remarkable presupposition, i.e. Theory of Evolution. Darwin collected data and made a hypothesis called "Natural Selection". Later as we kept digging we found more evidence of his hypothesis. However, he was not the first person to make that hypothesis as we have learnt in the first chapter. Today we can say with full confidence that Evolution is a fact, even though some educated idiots such as Ben Carson (the neurosurgeon with zero knowledge of neuroscience) with their baseless religious fundamentalism keep rambling otherwise. But in the beginning it lacked material evidence. So, it's not about the lack of evidence. Any new theory shall always lack material evidence and that's exactly what makes it a new theory. What matters is that, whether that hypothesis sounds plausible enough or it is way far-fetched like the pompous concepts of singularity and transhumanism. So, in simple terms, Science is always about the best guess based on available data.

If we live long enough, we can prove any plausible hypothesis to be true. But, explaining "Reality" is not as simple as digging out hominid fossils to back up the Theory of Evolution. It's much more than that. Technically, there is no such thing as reality. Because everything we perceive with our human consciousness is just a virtual simulation in our brain circuits. And this virtual simulation is constantly affected by the environment and various biological processes. The simplest example of such constantly changing reality comes from the most common female term "PMS", or Pre-Menstrual Syndrome. And the extreme form of PMS is medically called "PMDD" or Pre-Menstrual Dysphoric Disorder.

PMS is a part of female existence. If you are a woman, you already know what I'm talking about. But if you are a man, then imagine a reality that is never constant but rapidly changing from week to week. That's exactly what a woman faces in her life. The hormonal interplay inside a woman's head creates her reality. Her hormones tell her day to day what's important. They mold her desires and values. She goes through these stormy fluctuations of various hormonal states since her girlhood till menopause.

Due to this stormy weather inside a woman's head, her version of reality is never as constant as a man's. A man's mental universe is like the Himalayas standing millennia after millennia invincible to massive changes in geography, whereas the female universe is like the unpredictable climate change. A man can never even imagine in his wildest dreams how the storm inside the woman's head feels like. These storms often make a woman completely misunderstood by her man.

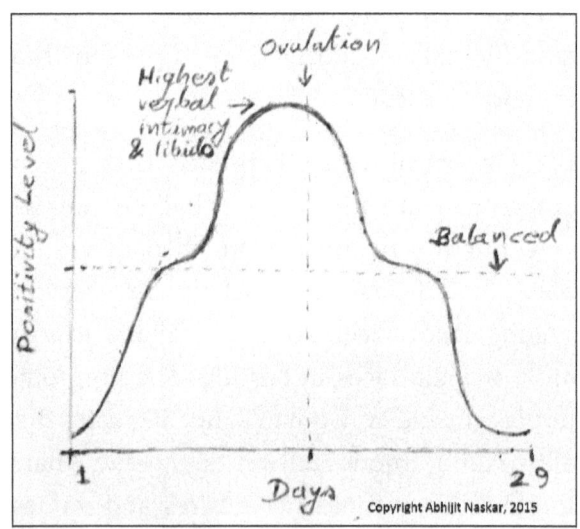

Figure 6-1: The Levels of Positive Cognitive Reality of a woman throughout the month

In all menstruating women, the female brain goes through various changes. Most weeks of the month women are brainy, creative, enthusiastic, cheerful and optimistic, but a mere shift in the hormonal flood on certain days turns their cheerful conscious reality into a gloomy one. On those days they tend to hate themselves and their lives. And the most fascinating thing about those days is that the hopelessness caused by hormonal imbalances feels so damn real to a woman that she literally perceives it as the everlasting reality of her life. The hormonal turbulence makes her absolutely blind to all the cheerful moments of her life. Once the hormonal storm wears off, the hopelessness and gloom fade away and she comes back to her original sunny state gaining her balanced natural conscious reality. This is a basic biological phenomenon in the life of every woman.

Changes in merely some hormonal tides completely alter the reality of a woman. For a woman it's all about feeling blue during those tides. A little anomaly in the circuits of various brain regions can make up a completely false reality for a person, such as in temporal lobe epilepsy. In this reality, one experiences visual and auditory hallucinations.

The perceived reality in temporal lobe epilepsy can be so convincing that it can make the percipient carry out some really absurd feats. The history of human civilization is filled with individuals suffering from temporal lobe epilepsy, such as Fyodor Dostoyevsky, Joan of Arc, and many more. Individuals with temporal lobe epilepsy often report vivid hallucinatory experiences of mystical nature. And a common element of temporal lobe epilepsy, is the vividly ecstatic encounter with God.

Recorded history of neurological evidence shows that transient electrical microseizures within the temporal lobes of the human brain, especially the right temporal lobe evoked vivid encounters with spiritual beings. By isolating the controlling stimuli that evokes a mystical experience today we are able to create similar experiences inside the laboratory. To the layman epileptic seizures mean someone convulsing and losing consciousness. But that's just one type of seizure and it is not even the most common kind. Many are just unaware of a whole other category of seizures, known as partial seizures that can cause a kaleidoscope of symptoms, such as the sense of oneness, complex hallucinations and feelings of fear, depression, and euphoria. Often, these seizures don't involve any convulsions at all and in some cases they can invoke the presence of God. During these God experiences, whatever messages the individual apparently receives from that illusory Supreme Being are just his own instinctual subconscious desires. During these microseizures an individual who craves for philosophical guidance would receive the gospel from his own instincts through the imaginary spiritual being, likewise a person with nomadic instincts would vividly hallucinate the promise of a new land.

Throughout history, nomads believing themselves to be "chosen" have marched into lands belonging to others and slaughtered them, all because of a vivid gut instinct: *"God gave us the land"*. Anomalous temporal lobe activity led to the arrival of the Ten Commandments. Temporal lobes are highly interconnected with the limbic system, especially the emotion and fear center of the brain, i.e. amygdala. So, when the temporal lobe acts anomalously, it sometimes triggers excitement in the amygdala, which often leads to a feeling of unearthly bliss and absolute meaningfulness of the universe. Such feeling often overwhelms an individual during the God encounter. Specific electrical anomalies in the amygdala can also trigger aggression and homicidal urges. This counts for the outrageous and murderous commands from man's God. Likewise, a person with insanely pervert subconscious desires, would justify those polygamous desires as the command of God.

"What we call rational grounds for our beliefs are often extremely irrational attempts to justify our instincts."

- Thomas Henry Huxley

The most significant literary contribution to this mystical phenomenon was made by the Russian Novelist Fyodor Dostoyevsky. He kept records of his 102 epileptic seizures during his last two decades, which mainly occurred at night. Seizures which occurred in the daytime were often preceded by an ecstatic aura. This has led neurologists to theorize that he had temporal lobe epilepsy. Based on his experiences, he created characters with epilepsy in his four novels The Possessed, The Brothers Karamazov, The Insulted and Injured, and The Idiot.

Dostoyevsky recorded his own experiences of temporal lobe epilepsy as:

"For several instants I experience a happiness that is impossible in an ordinary state, and of which other people have no conception. I feel full harmony in myself and in the whole world, and the feeling is so strong and sweet that for a few seconds of such bliss one could give up ten years of life, perhaps all of life. I felt that heaven descended to earth and swallowed me. I really attained god and was imbued with him. All of you healthy people don't even suspect what happiness is, that happiness that we epileptics experience for a second before an attack."

His writings are a treasure trove of blissful epileptic experiences for medical historians that clearly associates the experience of God with anomalous activity in the temporal lobe. In the year 1838 French psychiatrist Jean-Étienne Dominique Esquirol first recognized that temporal lobe epilepsy was the root cause of all kinds of spiritual/religious experiences.

Often temporal lobe epileptics report that God has given them a mission to transform the whole world. Just like Joan of Arc claimed that she had visions of the Archangel Michael, Saint Margaret and Saint Catherine instructing her to support Charles VII and recover France from English domination. Medical historians have shown that she suffered from tuberculosis and with a temporal lobe tuberculoma. A tiny brain lesion in her temporal lobe triggered her vivid visual and auditory hallucinations of various angels.

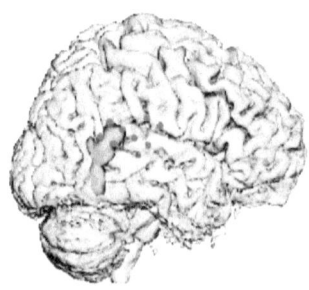

Figure 6-2: A modern day epileptic patient with lesions in the right temporal lobe leading to mystical experiences

The very reality of an omnipresent, omniscient and omnipotent being emerges from neural activity. For example, the right side of your brain is amazingly wired through millions of years to see the Big Picture, even when there is no actual picture. Right hemisphere draws the storyline of the spiritual/religious experience based on the person's fantasies, attitude, desires and beliefs. During anomalous temporal lobe functions, the brain simulates such a vivid conscious reality that the individual becomes convinced that he/she has accessed the cosmic consciousness of the universe. You can find the detail elaboration on the neuroscientific foundation of all kinds of religious/spiritual/mystical experiences in my book The God Parasite: Revelation of Neuroscience.

All conscious experiences such as, déjà vu, time distortion, sensed presences of God, angels, ghosts or aliens have been elicited in healthy individuals by stimulating various regions of the brain, especially the right temporal cortex in various neurological studies.

Brain circuits with various neurological conditions (fatal or harmless) can completely alter one's conscious perception of the reality. The perception of reality is often referred to as

"Qualia" in the neuroscientific community. Qualia (plural form of "quale") are the raw feelings of conscious experience such as, the taste of your lover's lips, smell of his/her body, the redness of red, the painfulness of pain etc. Synesthesia is such an extraordinary neurological phenomenon, where an individual's physical senses get mixed up. Sensations evoked through one sensory pathway produce vivid qualia normally associated with another physical sense. Perhaps it would be easier for you to understand if I mention the experiences of synesthesia. Individuals with this condition report physically 'seeing' sounds/smells/tastes or 'tasting' colors/sounds or 'hearing' colors and so on.

However some forms of synesthesia can occur during stages of meditation, sensory deprivation or even during the use of psychedelics such as LSD, marijuana etc. And the biological cause behind synesthesia is merely cross-connection between different brain regions of the somatosensory system involved in different sensory functions.

Conscious awareness of the reality is a mysterious thing. Cognitive reality of an individual solely arises from the make-up of the brain structure. Any kind of damage, like stroke can alter this reality without the awareness of the individual. Let me tell you a story of a stroke patient with such an altered cognitive reality. Her condition was noticed by my friend and colleague V.S. Ramachandran, one of the most interesting neuroscientists of our times. He explains her condition like this:

"Who was this rolling out of the bedroom in a wheelchair? Sam couldn't believe his eyes. His mother, Ellen, had just returned home the night before, having spent two weeks at the Kaiser Permanente hospital recuperating from a stroke. Mom had always been fastidious about her looks. Clothes and makeup were Martha Stewart perfect, with beautifully coiffed hair and fingernails painted in tasteful shades of pink or red. But today something was seriously wrong. The naturally curly hair on the left side of Ellen's head was uncombed, so that it stuck out in little nestlike clumps, whereas the rest of her hair was neatly styled. Her green shawl was hanging entirely over her right shoulder and dragging on the floor. She had applied rather

bright red lipstick to her upper right and lower right lips, leaving the rest of her mouth bare. Likewise, there was a trace of eyeliner and mascara on her right eye but the left eye was unadorned. The final touch was a spot of rouge on her right cheek —

very carefully applied so as not to appear as if she were trying to hide her ill health but enough to demonstrate that she still cared about her looks. It was almost as though someone had used a wet towel to erase all the makeup on the left side of his mother's face!

"Good grief!" cried Sam. "What did you do to your makeup?"

Ellen raised her eyebrow in surprise. What was her son talking about? She had spent half an hour getting ready this morning and felt she looked as good as she possibly could, given the circumstances.

Ten minutes later, as they sat eating breakfast, Ellen ignored all the food on the left side of her plate, including the fresh-squeezed orange juice she so loved.

Sam raced for the phone and called me (Ramachandran), as one of the physicians who had spent time with his mother at the hospital. Sam and

I had gotten to know one another while I had been seeing a stroke patient who shared a room with his mother. "It's all right," I said, "don't be alarmed. Your mother is suffering from a common neurological syndrome called hemi-neglect, a condition that often follows strokes in the right brain, especially in the right parietal lobe. Neglect patients are profoundly indifferent to objects and events in the left side of the world, sometimes including the left side of their own bodies."

"You mean she's blind on the left side?"

"No, not blind. She just doesn't pay attention to what's on her left. That's why we call it neglect."

The next day I was able to demonstrate this to Sam's satisfaction by doing a simple clinical test on Ellen. I sat directly in front of her and said, "Fixate steadily on my nose and try not to move your eyes." When her gaze was fixed, I held my index finger up near her face, just to the left of her nose, and wiggled it vigorously.

"Ellen, what do you see?"

"I see a finger wiggling," she replied.

"Okay," I said. "Keep your eyes fixed on the same spot on my nose." Then, very slowly and casually, I

raised the same finger to the same position, just left of her nose. But this time I was careful not to move it abruptly. "Now what do you see?"

Ellen looked blank. Without having her attention drawn to the finger—via motion or other strong cues—she was oblivious. Sam began to understand the nature of his mother's problem, the important distinction between blindness and neglect. His mother would ignore him completely if he stood on her left side and did nothing. But if he jumped up and down and waved his arms, she would sometimes turn around and look.

For the same reason, Ellen fails to notice the left side of her face in a mirror, forgets to apply makeup on the left side of her face, and doesn't comb her hair or brush her teeth on that side. And, not surprisingly, she even ignores all the food on the left side of her plate. But when her son points to things in the neglected area, forcing her to pay attention, Ellen might say, "Ah, how nice. Fresh-squeezed orange juice!" or "How embarrassing. My lipstick is crooked and my hair unkempt."

...

Finally, I took a sheet of paper, put it in front of Ellen and asked her to draw a flower.

"What kind of flower?" she said.

"Any kind. Just an ordinary flower."

Again, Ellen paused, as if the task were difficult, and finally drew another circle. So far so good. Then she painstakingly drew a series of little petals—it was a daisy—all scrunched on the right side of the flower (Figure 6-4). "

Figure 6-3: Drawing made by the hemineglect patient. Notice that the left half of the flower is missing. Many neglect patients will also draw only half the flower when drawing from memory—even with their eyes closed. This implies that the patient has also lost the ability to "scan" the left side of the internal mental picture of the flower. (Ramachandran, 1996)

Such neglect is not blindness, rather it is simply a general indifference to objects and events on the left. Let me give you another similar yet a little different kind of neglect case history.

A schoolteacher suffered a stroke that paralyzed the left side of her body, but she insists that her left arm is not paralyzed. Once, when she was asked, whose arm was lying in the bed next to her, she explained that the limb belonged to her brother. Then, when she was asked to clap, she proceeded to make clapping movements with her right hand, as if clapping with an imaginary left hand near the midline, while her left hand kept lying completely paralyzed with no movement whatsoever.

This patient was in fact completely paralyzed on the left side of her body after a stroke that damaged the right hemisphere of her brain. And like this one there is a small subset of patients with right hemispheric damage who seem to be absolutely unaware of the fact that the entire left side of their body is paralyzed even though they are quite mentally lucid in all other aspects. In the year 1908 French neurologist Joseph François Babinski first observed this curious disorder in which the patient's tendency is to ignore or sometimes even to deny the fact that one's left arm or leg is paralyzed. Babinski termed this condition as "Anosognosia" that means "unaware of illness".

Neglect stories are very popular in the field of neurology. There is another kind of neglect or denial in which a person downright denies to be alive. It is another fascinating neurophychiatric syndrome called 'Cotard' or 'walking corpse' syndrome.

This syndrome was named after the French neurologist Jules Cotard. He described the condition as 'le délire de negation' or 'the delirium of negation'. It results in a feeling that one is either dead or immortal. In 1880 Jules Cotard reported the case of a 43 year old lady, Mademoiselle X who believed that she had "no brain, nerves, chest or entrails and was just skin and bone", that "neither God nor the devil existed" and that "she was eternal and would live forever". The syndrome is described to have various degrees of severity, ranging from mild to severe. In a mild state, feelings of despair and self-loathing occur, whereas in the severe state the person with Cotard syndrome actually starts to deny the very existence of the self. In 2007 McKay and Cipolotti published a report on a 24 year old patient called LU. LU repeatedly thought that she was in heaven, even though she was actually in National Hospital, Queen Square, London and that she might have died from flu. The delusions diminished over a few days and were gone after a week.

Cotard delusion is usually associated with lesions in the parietal lobe as well as the prefrontal cortex. It can be treated with various antipsychotic, antidepressant and mood stabilizing drugs along with electroconvulsive therapy (ECT) and psychotherapy. The brain of an individual with Cotard syndrome, generates the conscious awareness of being dead. However such delusion does not hamper one's daily choirs and the person walks around and carries out daily activities just like a normal healthy person. The delusion goes away with treatment, but until it vanishes, it remains the only conscious reality to the person.

However, I must make something clear here. Individuals with these mental illnesses cannot be considered as "crazy", since they are completely lucid in all other daily activities.

These mental conditions are simply emergency defense measures constructed by the unconscious to deal with sudden overwhelming bewilderments about one's body and the space around it.

There is another mind-boggling neurological phenomenon, called Capgras' syndrome, where the patient sees familiar and loved figures as impostors. This delusion is one of the rarest and most colorful syndromes in neurology. The patient, who is often mentally quite lucid, comes to regard close acquaintances, usually his parents, children, spouse or siblings, as impostors. One patient reported with absolute belief: *"That man looks identical to my father but he really isn't my father. That woman who claims to be my mother? She's lying. She looks just like my mom but it isn't her."* Many of the documented cases of Capgras' syndrome have occurred in association with traumatic brain injury. This implies that the syndrome has a neurological basis.

Capgras' delusion results from a disconnection between the face recognition region in the temporal lobe and the emotion center of the brain, i.e. amygdala. Face recognition pathways remain completely normal, so a person with Capgras' could identify everyone, but as the communication between the face recognition region and amygdala is selectively damaged he/she would not experience any emotions when looking at the faces of his/her beloved ones. In the case of the patient mentioned earlier, he doesn't feel a "warm glow" when looking at his beloved mother, so when he sees her he says to himself, *"If this is my mother, why doesn't her presence make me feel like I'm with my mother?"* So the only way he could make sense of it, is to assume that this woman merely resembles his Mom, but is actually an impostor.

Often the brain of a person with Capgras' delusion creates some really bizarre cognitive reality. In one of such recorded case histories, a patient was convinced that his stepfather was a robot, proceeded to decapitate him and opened his skull to look for microchips.

But why exactly the close relatives are perceived as imposters and not any other familiar face? This is because when a person encounters someone who is emotionally very close to him/her, such as a parent, spouse or sibling, he/she naturally expects an emotional glow, a warm fuzzy feeling. The absence of this glow in the most expected relationship is therefore surprising which is then rationalized through an absurd delusion. On the other hand, when a person sees someone familiar but not emotionally close, he/she doesn't expect a warm glow and consequently there is no need for the brain to generate a delusion to explain the lack of warm and fuzzy feeling.

Figure 6-4: Random jumble of splotches. Gaze at this picture for a few seconds and you will eventually see a Dalmatian dog sniffing the ground mottled with shadows of leaves. Once the dog has been seen, it is impossible to get rid of it. Neurons in

the temporal lobes become altered permanently after the initial brief exposure, once you have "seen" the dog. (Tovee, Rolls and Ramachandran, 1996)

Observing the medical histories of various neurological syndromes is like observing human nature and human consciousness through a magnifying lens. They remind us of the overwhelming aspects of human silliness. They make us realize how easily our own mind can play tricks on us.

The human construct of the so-called reality is prone to self–deception. One way or another, we all are being deceived by our own mind. We always see what we want to see. Every moment, we create a new reality, and then the earlier reality loses its accountability (Figure 6-5).

Perception of reality emerges from the brain and dissolves in the brain. All your hopes, happiness, aspirations and inspirations rise from the intricate and enchanting neural firings in your brain. This very scientific reality may seem a little materialistic to some people, but I believe nothing can be more exhilarating than this. By looking closely at the various subtle functions of the brain we shall get closer to the mysterious mind of the universe. Observing the mind-boggling laws of the human mind we shall truly discover the laws of the universe.

Bibliography

Anton, S. C. Natural history of Homo erectus. American Journal of Physical Anthropology S37, 126-70 (2003)

Alemseged, Z., Spoor, F., Kimbel, W.H., Bobe, R., Geraads, D., Reed, D., Wynn, J.G., 2006. A juvenile early hominin skeleton from Dikika, Ethiopia. Nature 443, 296-30.

Asfaw, B., White, T., Lovejoy, O., Latimer, B., Simpson, S., Suwa, G., 1999. Australopithecus garhi: a new species of early hominid from Ethiopia. Science 284, 629-635.

Antón, S.C., 2003. Natural history of Homo erectus. Yearbook of Physical Anthropology 46, 126–170.

Armstrong E (1982) Mosaic evolution in the primate brain: differences and similarities in the hominoid thalamus. In: Armstrong E, Falk D (eds) Primate brain evolution: methods and concepts. Plenum, New York, pp 131–162

Baars, B. (1988), A Cognitive Theory of Consciousness (New York: Cambridge University Press).

Bancaud, J., Brunet-Bourgin, F., Chauvel, P. & Halgren, E. (1994), 'Anatomical origin of deja vu and vivid "memories" in human temporal lobe epilepsy', Brain, 117, pp. 71–90.

Bear, D.M. (1979), 'Personality changes associated with neurologic lesions', in Textbook of Outpatient Psychiatry, ed. A. Lazare (Baltimore, MD: Williams and Wilkins Co.).

Bogen,J.E.(1995a), 'On the neurophysiology of consciousness: Part I. An overview', Consciousness and Cognition, 4, pp. 52–62.

Bogen, J.E. (1995b), 'On the neurophysiology of consciousness: Part II. Constraining the semantic prob- lem', Consciousness and Cognition, 4, pp. 137–58.

Buxhoeveden DP, Switala AE, Litaker M, Roy E, Casanova MF (2001) Lateralization of minicolumns in human planum temporale is absent in nonhuman primate cortex. Brain Behav Evol 57:349–358

Bregman, A. (1981), 'Asking the ''what for'' question', in Perceptual Organization, ed. M. Kubovy & J. Pomerantz (Hillsdale, NJ: Lawrence Erlbaum Associates).

Blumenschine, R. J. et al. Late Pliocene Homo and hominid land use from Western Olduvai Gorge, Tanzania. Science 299, 1217-12121 (2003)

Brunet ,M., Guy, F., Pilbeam,. D., Mackaye, H.T., Likius, A., Ahounta, D., Beauvilain, A., Blondel, C., Bocherens, H., Boisserie, J.R., De Bonis, L., Coppens, Y., Dejax, J., Denys, C., Duringer, P., Eisenmann, V.R., Fanone, G., Fronty, P., Geraads, D., Lehmann, T., Lihoreau, F., Louchart, A., Mahamat, A., Merceron, G., Mouchelin, G., Otero, O., Campomanes, P.P., De Leon, M.P., Rage, J.C., Sapanet, M., Schuster, M., Sudre, J., Tassy, P., Valentin, X., Vignaud, P., Viriot, L., Zazzo, A., Zollikofer, C., 2002. A new hominid from the Upper Miocene of Chad, central Africa. Nature 418(6894), 145-151

Brunet, M., Guy, F., Pilbeam, D., Lieberman, D.E., Likius, A., Mackaye, H.T., de Leon, M.S.P., Zollikofer, C.P.E., Vignaud, P., 2005. New material of the earliest hominid from the Upper Miocene of Chad. Nature 434(7034), 752-755.

Berger, L.R., Clarke, R.J., 1995. Eagle involvement of the Taung child fauna. Journal of Human Evolution 29, 275-299.

Berger, T., Trinkaus, E., 1995. Patterns of trauma among the Neandertals. Journal of Archaeological Science 22, 841-852.

Berger, L.R., de Ruiter, D.J., Churchill, S.E., Schmid, P., Carlson, K.J., Dirks, P.H.G.M., Kibii, J.M., 2010. Australopithecus sediba: A New Species of Homo-Like Australopith from South Africa. Science 328, 195-204.

Bickerton, D. (2009). Adam's tongue: How humans made language and how language made humans. New York: Hill and Wang. Brothers, L. (2002). The social brain: A project for integrating primate behavior and neurophysiology in a new domain. In J. T. Cacioppo et al. (Eds.), Foundations in neuroscience, pp. 367. Cambridge, MA: MIT Press.

Balter, M., 2010. Candidate human ancestor from South Africa sparks praise and debate. Science 328, 154-155.

Bobe, R., Behrensmeyer, A.K., 2004. The expansion of grassland systems in Africa in relation to mammalian evolution and the origin of the genus Homo. Palaeogeography, Palaeoclimatology, Palaeoecology 207, 399-420.

Comparative primate energetics and hominid evolution. American Journal of Physical Anthropology 102, 265–281.

Clarke, R.J., Tobias, P.V., 1995. Sterkfontein Member 2 foot bones of the oldest South African hominid. Science 269, 521–524.

Churchland, P.S. (1986), Neurophilosophy (Cambridge, MA: The MIT Press). Churchland, P.S. (1996), 'The hornswoggle problem', Journal of Consciousness Studies, 3 (5–6), pp. 402–8.

Churchland, P.S. & Ramachandran, V.S. (1993), 'Filling in: Why Dennett is wrong', in Dennett and His Critics: Demystifying Mind, ed. B. Dahlbom (Oxford: Blackwell Scientific Press).

Churchland, P.S., Ramachandran, V.S. & Sejnowski, T.J. (1994), 'A critique of pure vision', in Large- scale Neuronal Theories of the Brain, ed. C. Koch & J.L. Davis (Cambridge, MA: The MIT Press).

Cicurel, R., "L'ordinateur ne digérera pas le cerveau", Sarina Editions, 2013

Cobb S., Ramachandran, V.S. & Hirstein, W. (in preparation), 'Evoked potentials during synesthesia'. Cohen, M.S., Kosslyn, S.M., Breiter, H.C. et al. (1996), 'Changes in cortical activity during mental rotation. A mapping study using functional MRI', Brain, 119, pp. 89–100.

Crick, F. (1994), The Astonishing Hypothesis: The Scientific Search for the Soul (New York: Simon and Schuster). Crick, F. (1996), 'Visual perception: rivalry and consciousness', Nature, 379, pp. 485–6.

Crick, F. & Koch, C. (1992), 'The problem of consciousness', Scientific American, 267, pp. 152–9.

Darwin, Charles. "On the origin of species by means of natural selection" (original edition, 1859).

Darwin, Charles. "The Descent of Man" (original edition, 1871).

Dawkins, R. "The Selfish Gene", Oxford University Press, 1976

Dawkins, R. "The Magic of Reality", Bantam Press, 2011

Dart, R.A. Australopithecus africanus: the southern ape-man of Africa. Nature 115, 195-199 (1925)

Dennett, D.C. (1991), Consciousness Explained (Boston, MA: Little, Brown and Co.).

Devinsky, O., Feldmann, E., Burrowes, K. & Broomfield, E. (1989), 'Autoscopic phenomena with seizures', Archives of Neurology, 46, pp. 1080–8.

Devinsky, O., Morrell, MJ, Vogt, BA. (1995) 'Contribution of anterior cingulate cortex to behavior', Brain, 118, pp. 279–306.

Domínguez-Rodrigo, M., Pickering, T.R., Semaw, S., Rogers, M.J., 2005. Cutmarked bones from Pliocene archaeological sites at Gona, Afar, Ethiopia: Implications for the functions of the world's oldest stone tools. Journal of Human Evolution 48, 109-121.

Dirks, P.G.H.M, Kibii, J.M., Kuhn, B.F., Steininger, C., Churchill, S.E., Kramers, J.D., Pickering, R., Farber, D.L., Mériaux, A.-S., Herries, A.I.R, King, G.C.P., Berger, L.R., 2010. Geological setting and age of Australopithecus sediba from Southern Africa. Science 328, 205-208.

DeGiorgio, M. et al. Out of Africa: modern human origins special feature: explaining worldwide patterns of human genetic variation using a coalescent-based serial founder model of migration outward from Africa. PNAS USA 106, 16057-16062 (2009)

Delson, E., Harvati, K., 2006. Return of the last Neanderthal. Nature 443, 762-763.

Dubois, E.,. 1894. Pithecanthropus erectus: eine menschenaehnlich Uebergangsform aus Java. Batavia: Landsdrukerei.

Edelman, G. M. (1992). Bright air, brilliant fire: On the matter of the mind. New York: Basic Books.

Enard W, Przeworski M, Fisher SE, Lai CS, Wiebe V, Kitano T, Monaco AP, Pääbo S (2002) Molecular evolution of FOXP2, a gene involved in speech and language. Nature 418:869–872

Falk, D. et al. Early hominid brain evolution: a new look at old endocasts. Journal of Human Evolution 38, 695-717 (2000)

Farah, M.J. (1989), 'The neural basis of mental imagery', Trends in Neurosciences, 10, pp. 395–9.

Fiorini, M., Rosa, M.G.P., Gattass, R. & Rocha-Miranda, C.E. (1992), 'Dynamic surrounds of receptive fields in primate striate cortex: A physiological basis', Proceedings of the National Academy of Science 89, pp. 8547–51.

Fodor, J.A. (1975), The Language of Thought (Cambridge, MA: Harvard University Press). Frith, C.D. & Dolan, R.J. (1997), 'Abnormal beliefs: Delusions and memory', Paper presented at the May, 1997, Harvard Conference on Memory and Belief.

Finlay BL, Darlington RB (1995) Linked regularities in the development and evolution of mammalian brains. Science 268:1578–1584

Gazzaniga, M. S. (1985). The social brain. New York: Basic Books. Greenspan, S. I. and S. G. Shanker (2004). The first idea: How symbols, language, and intelligence evolved from our early primate ancestors to modern humans. Cambridge, MA: Da Capo Press.

Gazzaniga, M.S. (1993), 'Brain mechanisms and conscious experience', Ciba Foundation Symposium, 174, pp. 247–57. Gloor, P., Olivier, A., Quesney, L.F., Andermann, F., Horowitz, S. (1982), 'The role of the limbic system in experiential phenomena of temporal lobe epilepsy', Annals of Neurology, 12, pp. 129–43.

Gloor, P. (1992), 'Amygdala and temporal lobe epilepsy', in The Amygdala: Neurobiological Aspects of Emotion, Memory and Mental Dysfunction, ed J.P. Aggleton (New York: Wiley-Liss).

Goodman M, Grossman LI, Wildman DE (2005) Moving primate genomics beyond the chimpanzee genome. Trends Genet 21:511–517

Goldberg, G., Mayer, N. & Toglis, J.U. (1981), 'Medial frontal cortex and the alien hand sign', Archives of Neurology, 38, pp. 683–6.

Grady, D. (1993), 'The vision thing: Mainly in the brain', Discover, June, pp. 57–66.

Green, R.E. A draft sequence of the Neandertal genome. Science 328, 710-722

Gilbert SL, Dobyns WB, Lahn BT (2005) Genetic links between brain development and brain evolution. Nat Rev Genet 6:581–590

Halgren, E. (1992), 'Emotional neurophysiology of the amygdala within the context of human cognition', in The Amygdala: Neurobiological Aspects of Emotion, Memory and Mental Dysfunction, ed J.P. Aggleton (New York: Wiley-Liss).

Hirstein, W. & Ramachandran, V.S. (1997), 'Capgras syndrome: A novel probe for understanding the neural representation of the identity and familiarity of persons', Proceedings of the Royal Society of London, 264, pp. 437–44.

Harcourt-Smith, W. E. & L.C. Aiello. Fossils, feet and the evolution of human bipedal locomotion. Journal of Anatomy 204, 403-416 (2004)

Haile-Selassie, Y., Suwa, G., White, T.D., 2004. Late Miocene teeth from Middle Awash, Ethiopia, and early hominid dental evolution. Science 303, 1503-1505.

Horgan, J. (1994), 'Can science explain consciousness?', Scientific American, 271, pp. 88–94.

Humphrey, N. (1993), A History of the Mind (London: Vintage).

Hublin, J.J. The origin of Neanderthals. PNAS 45, 169-177 (2009)

Henshilwood, C.S., Marean, C.W., 2003. The origin of modern human behavior: critique of the models and their test implications. Current Anthropology 44, 627-651.

Hof PR, Nimchinsky EA, Perl DP, Erwin JM (2001) An unusual population of pyramidal neurons in the anterior cingulate cortex of hominids contains the calcium- binding protein calretinin. Neurosci Lett 307:139–142

Hilton, C.E. (Eds) From Biped to Strider: The Emergence of Modern Human Walking, Running, and Resource Transport. Kluwer Academic/Plenum, New York, pp 50-52.

Haeusler, M., McHenry, H., 2004. Body proportions of Homo habilis reviewed. Journal of Human Evolution 46, 433-465.

Hobbs, J. (2006). The origins and evolution of language: A plausible strong-AI account. In M. Arbibi (Ed.), Action to language via the mirror neuron system. Cambridge: Cambridge University Press.

Holloway RL, Broadfield DC, Yuan MS (2004) The human fossil record, vol 3, Brain endocasts: the paleo- neurological evidence. Wiley, New York

Holloway RL (1996) Evolution of the human brain. In: Lock A, Peters CR (eds) Handbook of human symbolic evolution. Oxford University Press, Oxford, pp 74–114

Johanson, D.C., White, T.D., Coppens, Y. 1978. A new species of the genus Australopithecus (Primates: Hominidae) from the Pliocene of Eastern Africa. Kirtlandia 28, 2-14.

Johanson, D.C., Edey, M.E., 1981. Lucy: The Beginnings of Humankind. St Albans, Granada.

Jackendoff, R. (1987), Consciousness and the Computational Mind (Cambridge, MA: The MIT Press).

Jackson, F. (1986), 'What Mary did not know', Journal of Philosophy, 83, pp. 291–5.

Kanizsa, G. (1979), Organization In Vision (New York: Praeger).

Kuypers HGJM (1958) Corticobulbar connections to the pons and lower brainstem in man. Brain 81:364–388

Kinsbourne, M. (1995), 'The intralaminar thalamic nucleii', Consciousness and Cognition, 4, pp. 167–71.

Kimbel, W.H., Delezene, L.K., 2009. "Lucy" redux: A review of research on Australopithecus afarensis. Yearbook of Physical Anthropology 52, 2-48.

Kimbel, W. H. et al. Systematic assessment of a maxilla of Homo from Hadar, Ethiopia. American Journal of Physical Anthropology 103, 235-262 (1997)

Kunimatsu, Y. et al. A new Late Miocene great ape from Kenya and its implications for the origins of African great apes and humans. PNAS USA 104, 19661-19662. (2007)

King, W., 1864. The reputed fossil man of the Neanderthal. Quarterly Review of Science 1, 88-97.

Lackner,J.R.(1988),'Someproprioceptiveinfluenc esonperceptualrepresentations',Brain,111,pp.28 1–97.

LeDoux, J.E. (1992), 'Emotion and the amygdala', in The Amygdala: Neurobiological Aspects of Emo- tion, Memory and Mental Dysfunction, ed J.P. Aggleton (New York: Wiley-Liss).

Lalueza-Fox, C., Römpler, H., Caramelli, D., Stäubert, C., Catalano, G., Hughes, D., Rohland, N., Pilli, E., Longo, L., Condemi, S., de la Rasilla, M., Fortea, J., Rosas, A., Stoneking, M., Schöneberg, T., Bertranpetit, J., Hofreiter, M., 2007. A Melanocortin 1 Receptor Allele Suggests Varying Pigmentation Among Neanderthals. Science 318, 1453-1455.

Lacruz, R.S., Rozzi, F.R, Bromage, T.G., , 2005. Dental enamel hypoplasia, age at death, and weaning in the Taung child. South African Journal of Science 101, 567-569.

Le Gros Clark W.E., 1964. The fossil evidence for human evolution, 2nd ed. Chicago: University of Chicago Press. Leonard, W.R., Robertson, M.L., 1997.

Leakey, L.S.B., Tobias, P.V., Napier, J.R., 1964. A new species of the genus Homo from Olduvai Gorge. Nature 202, 7-9.

Lakoff, G. and M. Johnson (1999). Philosophy in the flesh. Basic Books: New York. LeDoux, J. E. (1996). The emotional brain. New York: Simon & Schuster.

Maryansky, A. (1996). African Ape social structure: A blue print for reconstructing early hominid structure. In J. Steel, S. Sherman (Eds.), The Archeology of Human Ancestry. London: Rutledge.

Massey, D. (2000). What I don't know about my field but wish I did. Annual Review of Sociology, 26(1), 699.

Massey, D. S. (2002). A brief history of human society: The origin and role of emotion in social life: 2001 presidential address. American Sociological Review, 67(1), 1–29.

Miller, B. D. (2007). Cultural anthropology, 4th ed. Boston: Allyn & Bacon.

Mayr, E., 1950. Taxonomic categories of fossil hominids. Cold Spring Harbor Symp Quant Biol 25, 109–118.

Martinez, I., Rosa, L., Arsuaga, J.-L. Jarabo, P., Quam, R., Lorenzo, C., Gracia, A., Carretero, J.-M., Bermúdez de Castro, J.M., Carbonell, E., 2004. Auditory capacities in Middle Pleistocene humans from the Sierra de Atapuerca in Spain. Proceedings of the National Academy of Sciences 101, 9976-9981.

Mounier, A., Marchal, F., Condemi, S. 2009. Is Homo heidelbergensis a distinct species? New insight on the Mauer mandible". Journal of Human Evolution 56, 219-246.

McHenry, H., 1998. Body proportions in Australopithecus afarensis and A. africanus and the origin of the genus Homo. Journal of Human Evolution 35, 1-22.

McHenry, H. M. Body size and proportions in early hominids. American Journal of Physical Anthropology 87, 407-431 (1992)

McBrearty, S., Brooks, A., 2000. The revolution that wasn't: a new interpretation of the origin of modern humans. Journal of Human Evolution 39, 453-563.

MacLean, P.D. (1990), The Triune Brain in Evolution (New York: Plenum Press).

MacKay, D.M. (1969), Information, Mechanism and Meaning (Cambridge, MA: The MIT Press).

Marr, D. (1982), Vision (San Francisco: Freeman). Medawar, P. (1969), Induction and Intuition in Scientific Thought (London: Methuen).

Milner, A.D. & Goodale, M.A. (1995), The Visual Brain In Action (Oxford: Oxford University Press).

Nagel, T. (1974), 'What is it like to be a bat?', Philosophical Review, 83, pp. 435–50.

Nash, M. (1995), 'Glimpses of the mind', Time, pp. 44–52.

Nicolelis, M. & Cicurel, R., "The Relativistic Brain: How it works and why it cannot be simulated by a Turing machine", Kioss Press, 2015

Nielson, J.M. & Jacobs, L.L. (1951), 'Bilateral lesions of the anterior cingulate gyri', Bulletin of the Los Angeles Neurological Society, 16, pp. 231–4.

Nimchinsky EA, Gilissen E, Allman JM, Perl DP, Erwin JM and Hof PR (1999) A neuronal morphologic type unique to humans and great apes. Proc Natl Acad Sci USA 96:5268–5273

Novembre, J., J. K. Pritchard and G. Coop (2007). Adaptive drool in the gene pool. Nature Genetics, 39, 1188.

Persinger, "'I would kill in God's name' role of sex, weekly church attendance, report of a religious experience and limbic lability" Perceptual and Motor Skills 1997.

Persinger "Experimental simulation of the God experience" Neurotheology 2003. Persinger, Corradini, Clement, Keaney, et al "Neurotheology and its convergence with neuroquantology" NeuroQuantology 2010.

Persinger, Koren and St-Pierre "The electromagnetic induction of mystical and altered states within the laboratory" Journal of Consciousness Exploration and Research 2010.

Persinger "Case report: A prototypical spontaneous 'sensed presence' of a sentient being and concomitant electroencephalographic activity in the clinical laboratory" Neurocase 2008.

Persinger and Saroka "Potential production of Hughlings Jackson's "parasitic consciousness" by physiologically-patterned weak transcerebral magnetic fields: QEEG and source localization" Epilepsy & Behavior 28 (2013).

Persinger. "The neuropsychiatry of paranormal experiences". J Neuropsychiatry Clin Neurosci 2001.

Persinger "Experimental Facilitation of the Sensed Presence: Possible Intercalation between the Hemispheres Induced by Complex Magnetic Fields" Journal of Nervous and Mental Disease 2002.

Pepperberg, I. (2008). Alex and me. HarperCollins: New York. Richardson, K.

(1999). The making of intelligence. London: Phoenix.

Paré, D. & Llinás, R. (1995), 'Conscious and preconscious processes as seen from the standpoint of sleep-waking cycle neurophysiology', Neuropsychologia, 33, pp. 1155–68.

Penfield, W.P. & Jasper, H. (1954), Epilepsy and the Functional Anatomy of the Human Brain (Boston, MA: Little, Brown & Co.).

Penfield, W.P. & Perot, P. (1963), 'The brain's record of auditory and visual experience: a final summary and discussion', Brain, 86, pp. 595–696.

Penrose, R. (1994), Shadows of the Mind (Oxford: Oxford University Press).

Penrose, R. (1989), The Emperor's New Mind: Concerning Computers, Minds and The Laws of Physics (Oxford: Oxford University Press).

Plum, F. & Posner, J.B. (1980), The Diagnosis of Stupor and Coma (Philadelphia: F.A. Davis and Co.).

Posner, M.I. & Raichle, M.E. (1994), Frames of Mind (New York: Scientific American Library).

Preuss TM, Caceres M, Oldham MC, Geschwind DH (2004) Human brain evolution: insights from micro- arrays. Nat Rev Genet 5:850–860

Purpura K.P. & Schiff, N.D. (1997), 'The thalamic intralaminar nuclei: a role in visual awareness', The Neuroscientist, 3, pp. 8–15.

Pickford, M., Senut, B., 2001. 'Millennium Ancestor', a 6-million-year-old bipedal hominid from Kenya - Recent discoveries push back human origins by 1.5 million years. South African Journal of Science 97, 22-22.

Pickford, M., Senut, B., Gommery, D., Triel, J., 2002. Bipedalism in Orrorin tugenensis revealed by its femora. Comptes Rendus Palevol 1, 191-203.

Ramachandran, V.S. (1993), 'Filling in gaps in logic: Some comments on Dennett', Consciousness and Cognition, 2, pp. 165–8.

Ramachandran,V.S.(1995a),'Fillingingapsinlogic :ReplytoDurginetal.',Perception, 24,pp.41-845.

Ramachandran, V.S. (1995b), 'Perceptual correlates of neural plasticity', in Early Vision and Beyond, ed. T.V. Papathomas, C. Chubb, A. Gorea and E. Kowler (Cambridge, MA: The MIT Press).

Ramachandran, V.S. (1995c), 'Anosognosia in parietal lobe syndrome', Consciousness and Cognition, 4, pp. 22–51.

Ramachandran, V.S. & Gregory, R.L. (1991), 'Perceptual filling in of artificially induced scotomas in human vision', Nature, 350, pp. 699–702.

Ramachandran, V.S., Rogers-Ramachandran, D. & Cobb, S. (1995), 'Touching the phantom limb', Nature, 377, pp. 489–90.

Ramachandran, V.S. and Blakeslee, S. (1999), Phantoms in the Brain: Probing the Mysteries of the Human Mind (William Morrow Paperbacks)

Richmond, B.G., Jungers, W.L., 2008. Orrorin tugenensis femoral morphology and the evolution of hominin bipedalism. Science 319, 1662-1665.

Rilling, J. K. (2006). Human and nonhuman primate brains: Are they allometrically scaled versions of the same design? Evolutionary Anthropology, 15, 65.

Rilling JK Human and nonhuman primate brains: are they allometrically scaled versions of the same design? Evol Anthropol 15:65–77

Relethford, J. H. Genetic evidence and the modern human origins debate. Heredity 100, 555-563 (2008)

Rightmire, G. P. Out of Africa: modern human origins special feature: middle and later Pleistocene hominins in Africa and Southwest Asia. PNAS USA 106, 16046-16050 (2009)

Rightmire, G.P. Homo in the Middle Pleistocene: Hypodigms, variation, and species recognition. Evolutionary Anthropology 17, 8-21 (2008)

Roebroeks, W. & P. Villa. On the earliest evidence for habitual use of fire in Europe. PNAS USA Epub ahead of print (2011)

Roth, G. and Dicke, U. Evolution of the brain and intelligence, TRENDS in Cognitive Sciences Vol. 9, No. 5, 2005

Rightmire, G.P., 1998. Human evolution in the Middle Pleistocene: the role of Homo heidelbergensis. Evolutionary Anthropology 6, 218-227.

Searle, John R. (1980), 'Minds, brains, and programs', Behavioral and Brain Sciences, 3, pp. 417–58.

Searle, John R. (1992), The Rediscovery of the Mind (Cambridge, MA: The MIT Press).

Semendeferi K, Lu A, Schenker N, Damasio H (2002) Humans and great apes share a large frontal cortex. Nat Neurosci 5:272–276

Strauss, E., Risser, A. & Jones, M.W. (1982), 'Fear responses in patients with epilepsy', Archives of Neu- rology, 39, pp. 626–30.

Schoetensack, O., 1908. Der Unterkiefer des Homo heidelbergensis aus den Sanden von Mauer bei Heidelberg. Leipzig: Wilhelm Engelmann.

Stringer, C.B., Trinkaus, E., Roberts, M.B., Parfitt, S.A., Macphail, R.I., 1998.The Middle Pleistocene human tibia from Boxgrove. Journal of Human Evolution 34, 509-547.

Scott, R. S., Ungar, P.S., Bergstrom, T.S., Brown, C.A., Grine, F.E, Teaford, M.F., Walker, A., 2005. Dental microwear texture analysis shows within-species diet variability in fossil hominins. Nature 436, 693-695.

Schmid, P., 2004. Functional interpretation of the Laetoli footprints. In: Meldrum, D.J.,

Senut, B. et al. First hominid from the Miocene (Lukeino Formation, Kenya). C. R. Acad. Sci. Paris, Sciences de la Senut, B., Pickford, M., Gommery, D., Mein, P., Cheboi, K., Coppens, Y., 2001. First hominid from the Miocene (Lukeino Formation, Kenya). Comptes Rendus De L Academie Des Sciences Serie Ii Fascicule a-Sciences De La Terre Et Des Planetes 332, 137-144.

Sherwood CC, Broadfield DC,Gannon PJ,Holloway RL, Hof PR (2003) Variability of Brocas area homologue in African great apes: implications for language evolution. Anat Rec 71A:276–285

Spoor, F., Leakey, M.G., Gathogo, P.N., Brown, F.H., Antón, S.C., McDougall, I., Kiarie, C. Manthi, F.K, Leakey, L.N., 2007. Implications of new early Homo fossils from Ileret, east of Lake Turkana, Kenya. Nature 448, 688–691.

Stringer, C.B., Finlayson, J.C., Barton, R.N.E, Fernández-Jalvo, Y., Cáceres, I., Sabin, R.C., Rhodes, E.J., Currant, A.P., Rodríguez-Vidal, J., Giles-Pacheco, F., Riquelme-Cantal, J.A., 2008. Neanderthal exploitation of marine mammals in Gibraltar. Proceedings of the National Academy of Sciences USA 105, 14319–14324.

Shipman, P., 2008. Separating "us" from "them": Neanderthal and modern human behavior. Proceedings of the National Academy of Sciences USA 105, 14241-14242.

Schmitt, D., Churchill, S., 2003. Experimental evidence concerning spear use in Neandertals and early modern humans. Journal of Archaeological Science 30, 103-114.

Sutherland, N.S. (1989), The International Dictionary of Psychology (New York: Continuum).

Sawer, G. and Deak, V. (2007). The last human (p. 103). New York: Peter N. Nevraumont Publication – Yale University Press.

Small, D. (2008). On the deep history of the brain. Berkeley: University of California Press.

Turner, B. (2000a). Embodied ethnography. Doing culture. Social Anthropology, 8(1), 51.

Turner, J. H. (2000b). On the origins of human emotions: A sociological inquiry into the evolution of human affect. Stanford, California: Stanford University Press.

Tovee, M.J., Rolls, E.T. & Ramachandran, V.S. (1996), 'Rapid visual learning in neurones of the primate temporal visual cortex', Neuroreport, 7, pp. 2757–60.

Tulving, E. (1983), Elements of Episodic Memory (Oxford: Clarendon Press).

Trimble, M.R. (1992), 'The Gastaut-Geschwind syndrome', in The Temporal Lobes and the Limbic Sys- tem, ed. M.R. Trimble and T.G. Bolwig (Petersfield: Wrightson Biomedical Publishing Ltd.).

Trinkhaus, E., 1985. Pathology and the posture of the La Chappelle-aux-Saints Neanderthal.

American Journal of Physical Anthropology 67, 19-41.

Trinkaus, E., Shipman, P., 1993. The Neanderthals: Changing the Image of Mankind. Knopf: New York.

Thorpe, S.K.S., Holder, R.L., Crompton, R.H., 2007. Origin of human bipedalism as an adaptation for locomotion on flexible branches. Science 316, 1328-1331.

Terre et des planètes / Earth and Planetary Sciences 332, 137-144 (2001)

Ungar, P.S., Grine, F.E., Teaford, M.F., El-Zaatari, S., 2006. Dental microwear and diets of African early Homo. Journal of Human Evoution 50, 78–95

Ungar, P.S., Grine, F.E., Teaford, M.F., 2006. Diet in early Homo: a review of the evidence and a new model of adaptive versatility. Annual Review of Anthropology 35, 209-228.

Waxman, S.G. & Geschwind, N. (1975), 'The interictal behavior syndrome of temporal lobe epilepsy', Archives of General Psychiatry, 32, pp. 1580-6.

Ward, C. V. et al. Complete fourth metatarsal and arches in the foot of Australopithecus afarensis. Science 331, 750-753 (2011)

Ward, C. V. Interpreting the posture and locomotion of Australopithecus afarensis: where do we stand? American Journal of Physical Anthropology S35, 185-215 (2002)

White, T. D. et al. Ardipithecus ramidus and the paleobiology of early hominids. Science 326, 75-86 (2009)

Wong, K., 2010. Spectacular South African skeletons reveal new species from murky period of human evolution. Scientific American 8 April 2010

Wong, K., 2010. Fossils of our family. Scientific American June 2010.

Wray,A.(1998)."Protolanguageasaholisticsyste mforsocialinteraction,"Language & Communication 18, pp. 47–67.

Young, N. M. et al. The phylogenetic position of Morotopithecus. Journal of Human Evolution 46, 163-184 (2004)

Zeki, S.M. (1978), 'Functional specialisation in the visual cortex of the rhesus monkey', Nature, 274, pp. 423–8.

Zeki, S.M. (1993), A Vision of the Brain (Oxford: Oxford University Press).

Zollikofer, C. P. et al. Virtual cranial reconstruction of Sahelanthropus tchadensis. Nature 434, 755-759 (2005)

www.ingramcontent.com/pod-product-compliance
Lightning Source LLC
Chambersburg PA
CBHW030744180526
45163CB00003B/908